Transparent Designs

STUDIES IN COMPUTING AND CULTURE

This series publishes interdisciplinary, historically grounded works that examine our digital world in its broader sociocultural, political, economic, legal, and environmental contexts. With a global scope, it focuses on the relationships of digital technology to political economies and socioeconomic power inequalities and imbalances.

Transparent Designs

Personal Computing and the Politics of User-Friendliness

Michael L. Black

JOHNS HOPKINS UNIVERSITY PRESS BALTIMORE

Printed in the United States of America on acid-free paper
9 8 7 6 5 4 3 2 1

Johns Hopkins University Press
2715 North Charles Street
Baltimore, Maryland 21218-4363
www.press.jhu.edu

Library of Congress Cataloging-in-Publication Data

Names: Black, Michael L., 1984– author.
Title: Transparent designs : personal computing and the politics of user-friendliness / Michael L. Black.
Description: Baltimore : Johns Hopkins University Press, 2022. | Series: Studies in Computing and Culture | Includes bibliographical references and index.
Identifiers: LCCN 2021022699 | ISBN 9781421443539 (hardcover ; acid-free paper) | ISBN 9781421443546 (ebook)
Subjects: LCSH: User interfaces (Computer systems)—History—20th century. | Computer software—Human factors—History—20th century.
Classification: LCC QA76.9.U83 B49 2022 | DDC 005.4/37—dc23
LC record available at https://lccn.loc.gov/2021022699

A catalog record for this book is available from the British Library.

Portions of chapter 4 have been previously published in Michael L. Black, "Usable and Useful: On the Origins of Transparent Design in Personal Computing," *Science, Technology, & Human Values* 45, no. 3 (2020): 515–37. Copyright © 2020 Michael L. Black.

Contents

Acknowledgments

This book would not have been possible without the excellent mentoring that I received at the University of Illinois. I developed many of the core arguments I present in this book while writing my dissertation under the guidance of Robert Markley, Melissa Littlefield, Tim Newcomb, Spencer Schaffner, and Ted Underwood. Bob's work on the culture of science served as a model for my own research. His generous but always honest feedback on my writing also helped me to develop confidence as a scholar and feel comfortable reading my own work critically. Additionally, his advice on navigating academic publishing of all sorts has continued to prove invaluable. Melissa, Spencer, and Ted in different ways each helped me to refine my identity as an interdisciplinary scholar and to understand the relationship between the criticism and practice of technology, and Tim's feedback helped me to stay grounded in the humanities.

I would not have begun this project at all were it not for the encouragement that I received from mentors at Temple University. Miles Orvell, Susan Wells, and Peter Logan all taught me important lessons about how to navigate academia. Each encouraged me to follow my interest in studying digital culture and provided me with a space in their seminars to explore many of the ideas that set me on the path toward this project. Some of those ideas are ones I still think about regularly, and many are present in some form in this book. I could not have made it this far without the crucial support and mentoring they offered during the early stages of my academic career.

Writing a book is a long and difficult undertaking. In addition to support for the work of putting pen to paper, I also received advice and guidance on the planning, revision, and proposal process from a variety of people. Thank you, especially, to Jennifer Lieberman for your early feedback on my ideas and for your insights on how to communicate the importance of projects across different audiences. While this book began as a dissertation, no two words are the same. Most of the project's reenvisioning and restructuring occurred after I joined the faculty at the University of Massachusetts Lowell.

I could not have completed this book without advice and encouragement from my colleagues there. I would especially like to thank Anthony Szczesiul and Sue Kim for their feedback on early stages of the book as I rethought its style and presentation. Natalie Houston, Jonathan Silverman, and Todd Tietchen also gave me invaluable advice at other stages of the process. Thank you to all of my colleagues in the English Department for making me feel welcome and for supporting me in other ways as I have made my way through academia.

I would also like to thank Scott Poole and Kevin Franklin for giving me the opportunity to work full time as a digital humanist. My time at the Institute for Computing in the Humanities, Arts, and Social Sciences continues to influence my thinking on the various interests at stake in technology. While there, I worked on projects with faculty from all over the University of Illinois' campus. Those projects allowed me to see how other disciplines view computation, a challenging but unique experience that still informs my thinking about the complex cultural work that software performs and the network of influences that inform its design. Thank you, too, to Michael Simeone for introducing me to Scott, Kevin, Alan Craig, and others that I worked with while at the institute.

Early drafts of material from this book were presented at the Modern Language Association's annual conference, the Conference on College Composition and Communication, the Thomas R. Watson Conference on Rhetoric and Composition, and the Society for Literature, Science, and the Arts' annual conference. Thank you to the conferences' organizers for giving me a forum to discuss my work and to those in attendance for the questions, comments, and conversations that sometimes carried over into the hallways and over email after the end of sessions.

Archival research on early personal computing like the kind found in this book would have not been possible without the efforts of Jason Scott, the Bitsavers, Archive Team, and the Internet Archive. When I first began working on the research that would become a part of this book, I was digging through dusty old magazines in the labyrinthian stacks of Illinois' main library. After to moving to Massachusetts, I continued this effort, traveling around to other libraries in the Boston area. However, I eventually discovered that many of the sources I was reviewing had been scanned and made available online. The expanded version of this project was made possible through your work. I hope that other scholars are able to recognize the importance of the work you are doing.

Thank you to my parents, David and Debbie. Although I started research on this book while in graduate school, I have been thinking about many of the ideas in it since I was a kid. You never hesitated to encourage me to explore my interests in both computers and writing. I would not have made it this far without the strong foundation you provided for me. I also could not have written this book without Katherine's support. In moments of doubt or exhaustion, your support always helped me to find the energy to keep going.

Finally, I would like to thank the editors and staff at Johns Hopkins University Press who helped bring this book into print. My anonymous reviewers helped me to make many important decisions and to understand better my own research goals. Thank you to MJ Devaney for her help copyediting the manuscript. And finally, thank you especially to Matt McAdam for his advice and guidance throughout this process.

Transparent Designs

Introduction

The Politics of User-Friendliness

When asked in a 1984 interview to describe how the Macintosh would change personal computing, Steve Jobs responded by explaining that the goal of his team had been to "reach the point where the operating system is totally transparent. When you use a Lisa or a Macintosh, there is no such thing as an operating system. You never interact with it; you don't know about it."[1] Given the emphasis that Apple would place on the visuality of the Macintosh's operating system in its advertising campaigns, Jobs's remarks seem odd. Recent accounts of the Macintosh's design characterize Jobs as obsessive about managing even the most minute details of its graphic user interface, with one biographer noting that Jobs went so far as to insist on specific adjustments to the design of something as seemingly banal as a calculator application.[2] Technology journalists have attributed the fact that the Macintosh was able to survive commercially alongside other early attempts at commercial graphic user interfaces to Apple's development of and support for graphically rich software. The Macintosh was celebrated during the late 1980s and early 1990s for introducing typography to word processing, supporting real-time graphic design, and facilitating a growing desktop publishing industry. Even as the same or equivalent applications became available on other operating systems, the Macintosh maintained its reputation as the personal computer best suited for creative work.[3] In light of the varying roles that graphic features have played in the Macintosh's design and legacy, it seems strange that anyone, let alone its chief designer, would consider invisibility to be its main contribution to the history of computing. However, if we step back from the Macintosh itself, Jobs's remark is hardly surprising. As I show in this book, Jobs was participating in an ongoing discussion within the nascent personal computing industry about what counted as "user friendly."

Since the 1970s, outspoken proponents of a "personal" model of computing have argued that many of the complex social problems surrounding computers can be solved by simply putting the machines into the hands of as many people as possible. A "personal computer revolution" would decentralize computation, empowering ordinary people to explore as yet unknown cultural practices. This narrative framing has shaped our view of the first decade of personal computing in the United States, and it continues to play an important role in Silicon Valley's maintenance of its cultural authority. The constant pursuit of new technologies that are still somehow more revolutionary, that empower still more people in still more ways, sustains this narrative. Moments of doubt, uncertainty about the politics of Big Tech, and other critical responses to Silicon Valley's master narrative of its innovative benevolence are simply treated as evidence that the revolution is still ongoing.

This pattern is almost as old as personal computing itself. As sales of personal computers began to rise dramatically in the early 1980s, many industry participants and observers declared that the personal computer revolution had arrived. But along with increased access came concerns about the difficulty of developing widespread computer literacy and doubts that anyone other than a handful of specialists would find computers truly useful. Hobbyists, entrepreneurs, and engineers believed that computing would never be democratic so long as specialized knowledge remained a barrier to access, and so they turned their attention to questions of usability. They called for a new approach to design that would make computers more user friendly by hiding complicated, technical representations of computation and instead offer simpler, intuitively understandable interfaces. Computers would need to be "transparent" in a phenomenological sense so that users would not think of themselves as operating a machine at all. In pushing for transparency, they set us on a course towards the "friendly" future we now live in. Today's interfaces are designed to keep our focus on what we would like to do next, while the automated mechanisms beneath their surfaces handle the overly technical side of things for us, ostensibly on our behalf.

Usability has played and continues to play a critical role in a variety of fields. Throughout this book, I use the term to refer to theories and practices related to defining the usableness and usefulness of a given technology: how something should be used and what value is derived from its use. Several frameworks, including human-computer interaction, user experience design, and human-centered design, are available to today's usability experts.[4] Each of these frameworks has different emphases, but they all draw on foundational principles realized in the late 1970s and early 1980s through the design of commercial technologies and then codified in academic literature. During this period, usability experts working in personal computing advocated for transparent design by arguing that there was no benefit to requiring the average person to develop an in-depth understanding of the machines they would operate. Instead, they proposed interfaces that would represent computing in "friendlier," more familiar, and ideally intuitive terms.

Yet users' needs have never been the sole or even primary influence on design. Especially within the context of personal computing, usability is often normative in practice. In other words, we need to recognize that user-friendly design often serves to advance the designers' needs. For example, determining whether a certain use of copyrighted material is subject to the fair use exemption can be difficult. It is simpler not to require users to make that judgment call at all by placing significant restrictions on access and presenting users with an unintimidating interface that makes purchasing single-user licenses seem easy. Designing an interface that encourages users to browse and discover new content shifts their attention away from reflecting on their inability to see the mechanisms that surveil and constrain their behavior. And when we do notice them, we are encouraged to view them as acceptable, necessary, or even desirable.[5] In this sense, any discussion of usability is always about more than simply ease of use. Usability is always political in nature, a negotiation between the needs of a technology's designers and administrators and those of its users. In the chapters that follow, I explore conversations that unfolded in the late 1970s and early 1980s around usability to shed light on how transparency came to define the concept of user-friendliness in personal computing hardware and software design.

The Social Construction of User-Friendliness

Popular histories of personal computing often identify the release of the Macintosh as the moment personal computers finally were able to empower ordinary people because it was the first machine to make graphic user interfaces (GUIs) relatively affordable. However, narratives that position GUIs as the culmination of personal computing as a revolutionary movement obscure the various definitions of user-friendliness and usability that preceded them. The process through which transparent design became closely associated with user-friendliness in American computing culture extends far beyond the influence of Apple, Jobs, or the Macintosh. As Michael S. Mahoney and Nathan Ensmenger have noted, most published work on the history of personal computing has focused on documenting technologist "firsts."[6] Mahoney's and Ensmenger's critiques are directed at those academic histories that prioritize documenting lineages of materially influential technologies over representing how computers were viewed in American culture.[7] Alongside these histories are popular accounts written by journalists like Paul Freiberger, Michael Swaine, Steven Levy, and John Markoff that since the mid-1980s and early 1990s have helped shape popular understandings of digital culture in the United States.[8] While their work does better in tracking the discourse surrounding early personal computers, they have also uncritically portrayed personal computing as a politically engaged, countercultural movement. Despite suggesting that these new machines would be tools for liberation, they offer little in the way of specifics about how personal computers could contribute to efforts to undo the structural inequities that American activists had been protesting since the 1950s and that some believed had been made worse by computers.[9] Recently, many scholars have begun to raise serious doubts about their accounts.[10] Nonetheless, their portrayals of early personal computing continue to inform popular culture, where computers are still largely treated as tools for personal empowerment and where the political prognostications of a handful of hackers from the 1970s continue to be taken at face value. Although it is true that personal computing offered a model of usability that contrasted sharply with the norms of tightly centralized control and conformity associated with IBM, these popular histories today support the status quo vision of computing, serving as foundational texts that underwrite Silicon Valley's promotion of technological innovation as a primary driver of social and political progress.[11] As I argue in the chapters ahead, user-friendliness is a key part of this master narrative. Each reminder that our computers now "just work" serves to sustain a belief that Silicon Valley's unique perspective and entrepreneurial spirit are capable of leveraging computation to solve just about any problem we may face and even improve things we did not realize needed improving.

In contrast to these histories, my goal is not to identify a new set of firsts nor provide a more definitive account of the development and implementation of the specific machines I examine. Rather, I seek to understand how and why the belief that user-friendliness can only be achieved through transparent design has remained so durable in American digital culture. One key assumption supporting my work is that digital culture cannot be understood independently of the specific hardware and software we use to access it. In this respect, I believe it is important to expand our understanding of the rhetorical history of user-friendliness by considering the messaging and machines that preceded the Macintosh. Many talking points that contributed to the widespread adoption of transparency as a singular standard for user interface design emerged out of conversations about command-line-driven software. Stepping back from a focus on the development of GUIs allows us to see that public understandings of what it means for computers to be usable and useful were first formed in response to an array of early personal computers that each promised a distinct user experience despite having relatively similar models of usability. Many of the concepts we now associate with user-friendliness were first introduced by personal computers like the Apple II, Tandy TRS-80, Commodore PET, and IBM PC. Indeed, even were we to accept the idea that user-friendly computing begins with the Macintosh, then we would at least need to recognize that Apple emphasized the importance of its design by contrasting it with other, non-GUI-based machines that had been described previously as "user friendly." With the exception of the Apple II and to a lesser extent the IBM PC, the conversations hobbyists, businesspeople, designers, and programmers were having about these machines are not well represented in histories of personal computing. Understanding how and why transparency has come to displace other approaches to user-friendliness means returning to and engaging with past visions of our computational future that have been now forgotten in our GUI-focused constructions of digital culture.

Understanding the rhetorical history of user-friendliness also requires us to recognize that the first decade of computing was full of anxiety and concern about whether American manufacturers could realize the bold promises of a computer revolution that appeared in their advertisements.

To distinguish between the ways that computer use is discussed and the specific behaviors supported by a given computer system, I refer throughout this book to "rhetorics of usability" and "models of usability," respectively. Although advocates of transparent design suggest that interfaces are immediately usable if they are built around intuitively understandable signifiers, it is important to recognize that any sense of intuitiveness or immediate recognition of function is influenced by the ideas we internalize about how computers should work. Many of these ideas are often external to the specific interfaces we are interacting with. Past experiences with other devices, the written or digitally published material accompanying a device, public discussions of the uses for computers, and representations of computers in popular media all push us toward and away from a designer's intended interpretation of their model of usability. Many early efforts to promote user-friendliness, as I show, took place within advertising and public relations campaigns, whose goal was to reframe potentially controversial design decisions as made in pursuit of easier-to-use computers. If a model of usability represents a configuration of material affordances and constraints constructed to support an intended set of use behaviors, then a rhetoric of usability functions as a kind of top-level interface that mediates our interpretation of that configuration. Usability literature often focuses on the design and development of models of usability, but we must recognize that rhetorics of usability play an equally important role in normalizing user behavior by promising increased utility if users behave the way designers intend. User-friendliness encourages us to prioritize a grossly functional evaluation of technology. The only thing that matters is whether a given application works as promised, whether we believe that its model of usability fulfills the expectations set by the rhetorics of usability associated with it.

Importantly, user-friendliness also privileges models of universal usability, promising that well-designed technologies can be immediately usable by and substantially useful to everyone. Ruha Benjamin has observed, however, that universal approaches to design are often conducted from an "unmarked" perspective that ignores the different embodied experiences of users in favor of a bland white maleness. Universally usable technologies may feel alienating to or demand submission from those whose identities and experiences are not similarly unmarked.[12] Transparent design complicates matters further by potentially concealing those mechanisms whereby a software developer explicitly structures user agency and identity within a digital environment. The cultural, social, and political biases of the designer are often only felt by users because friendly interfaces do not offer a language through which to acknowledge these prejudices beyond vaguely reassuring error messages encouraging users to try again.[13]

My goal in tracing the rhetorical history of user-friendliness is to document how transparency's claims of intuitive and universal usability came to displace other approaches. In the chapters ahead, I examine user-friendliness in what Bruno Latour calls the "state of crisis in which machines, devices, and implements [are] born." During this state of crisis, the social ties and networks of influence that produce scientific theories and technological standards are visible because success is not yet assured. The uses and benefits of unproven technologies must be explained by their creators; and in the absence of established norms of practice, potential users and other stakeholders are more likely to receive substantive answers when they demand explanations. Although this moment has passed for personal computing, Latour suggests that it is possible to bring "back to light" the networks of influence that produced a technology "using archives, documents, memoirs, museum collections, etc. to artificially produce . . . the state of crisis."[14] I interpret Latour's suggested methodology here as one that encourages us to rely on archival material produced before, contemporaneously with, or just after the emergence of technologies we wish to study. My archive thus consists of the print culture circulating around commercial technologies, including personal computing magazines, users group newsletters, reference manuals, and advertisements, as well as articles appearing in general audience newspapers and news magazines published in the United States between 1974 and 1984.

Rhetorics of usability can also be subversive if they encourage users to adopt a different understanding of what counts as usable and useful. By tracing the development of a consensus around user-friendliness and transparent design, my goal is to develop an alternative rhetoric of usability that values complex engagements with technology rather than the simplified encounters manufactured in the name of ease or convenience. I am not alone in this regard. Sasha Costanza-Chock and Natasha N. Jones, for example, foreground principles of justice in design by arguing that technologies can only support the agency of users if they enable them to critique and intervene in models of usability.[15] That they feel compelled to issue calls for justice-based approaches to design is evidence that the boundary between designer and user that Lucy Suchman described in the early 1990s is very much still in place.[16] As I show, user-friendliness and transparent design emerged as justifications for this boundary through arguments to the effect that a benevolent oligarchy was necessary to address fundamental cultural, social, and technical issues that they believed held back the development of early personal computing technologies.

To trace the growing acceptance of this argument, I document the various concepts, rhetorical strategies, and narrative framings that journalists and other industry participants deployed to understand the sociocultural aspects of personal computing. While their ideas may not be as robust or rigorously defined as the academic discourse this book also engages with, I recognize that they served as theories of media, rhetoric, and literacy in their own right for them. Through these lenses, they interpreted and contributed to ongoing arguments about how and why personal computers should or should not be integrated into American society. Of course, these perspectives are not without their problems. Many uncritically assume technology can solve the very problems it creates, a belief that Meredith Broussard calls "technochauvinism," or betray a deeper commitment to a more traditionalist politics than their inclusive rhetoric might suggest via their embrace of what David Golumbia calls "computationalism."[17] In documenting the competing and conflicting social constructions of user-friendliness during the first decade of personal computing, I show how discussions of usableness and usefulness were about much more than simply making computers "easy to use." The rhetorics of usability supporting user-friendliness and transparent design often serve as a proxy for discussions about the politics of personal computing and more specifically as a justification for technologists to assume primary control over shaping the contexts that computers are invisibly integrated into.

Interrogating the Politics of User-Friendliness

The questions at the core of this book engage with research across a variety of fields. My own suspicion toward the politics of user-friendliness has been strongly influenced by research in digital media studies by Wendy Hui Kyong Chun, Lori Emerson, Alexander R. Galloway, N. Katherine Hayles, Lev Manovich, and Noah Wardrip-Fruin, among others, who have each demonstrated in different ways that the comparatively simple representations provided to us through transparent interfaces often strategically misrepresent the complex systems beneath them.[18] However, scholarship in digital media studies has largely been focused on genres of software and types of code used to create video games, electronic literature, and digital art that are valued for the novel critical and aesthetic work they perform. Like Robert Johnson, I believe that so-called mundane genres of software are important for understanding the politics of computing, even if they do not lend themselves as readily to the kinds of close reading common within the humanities.[19] More recently, Matthew Fuller and Andrew Goffey suggest that most of the software we use is a kind of "gray media" that is infrastructural in nature. We understand that these applications do not appear to be "the bearers of some sort of hidden meaning (they need not be the object of hermeneutics)" and that they possess "an unremarkableness that can be of inestimable value in background operations."[20] This intentional unremarkableness is something we should begin viewing with suspicion rather than continuing to welcome as a convenience.

Whereas more aesthetically complex genres of software conceal their mechanisms in order to call attention to the immersive experiences rendered on screen, these gray media are designed to fade from our awareness so that we can focus instead on the supposedly externally defined tasks that software merely supports. For example, scanning the interface of Microsoft Word, as Fuller does, to analyze the language embedded in its iconography or the "tooltips" that are displayed when holding a cursor over portions of it can help us understand how Word encourages us to view writing through the lens of Microsoft's corporate culture.[21] But the interface is only one source of influence on our ability to interpret and engage with digital culture. Focusing our attention on Microsoft's articulation of corporate culture via Word's interface shifts our attention away from the automated mechanisms beneath it that allow Word to function as an extension of the cloud-based surveillance tools that Microsoft provides employers access to via institutional Office365 subscriptions.[22] As I argue, rhetorics of usability that privilege user-friendliness and transparent design play an important role in keeping media gray by devaluing complex engagements with computation. One important goal of this book, then, is to extend the important questions that digital media scholarship raises in regards to the politics of computing into discussions of more workaday software.

Much of the distrust of transparent interfaces in digital media studies can be traced to Friedrich Kittler's problematization of layered software architecture. Kittler characterizes the relationships between the different layers of software that structure contemporary systems as built on "one-way functions" that "hide an algorithm from its result."[23] Computer systems, he explains, are designed by engineers to function as a series of "layers." While the lowest layer is hardware, each successive layer is comprised of software that remediates those below it, providing a new, distinct representation of the system's state and function. This process occurs first via the translation of circuitry into hardware addresses and continues with each new software layer until users have only the interface available to them to understand the machine before them.[24] Yet Kittler also observes that the function of each successive layer is bounded by the layers below it. In this sense, our relationship to computer systems as users is defined by the obscured layers that the interface rests atop as much as, if not more so, than by the interface itself. Ultimately, however, we can only understand those relationships in the terms that software developers provide to us via the interface, its documentation, or related promotional materials that describe its function. If we understand that software is always an incomplete representation of the algorithmic systems supporting it, then documenting and interrogating the specific choices made to hide or reveal certain aspects of computation allows us to understand the various interests that transparent design serves, which, in turn, can better help us to recognize not only why this approach to user-friendliness has remained so durable but also how it continues to support particular narratives of technological power in contemporary American culture.

Given their suspicion of interfaces, digital media scholars have developed a variety of methods for studying the formal elements of software that do not rely solely on the interface. Many of these methods were proposed in response to Nick Montfort's call to avoid engaging in "screen essentialism."[25] Matthew G. Kirschenbaum's "forensic criticism" and N. Katherine Hayles' "media-specific criticism" are two methods for describing software's meaning-making processes that begin with an awareness of the material construction of software and that read screenic representations against the processes that support it.[26] Elsewhere, "critical code studies" and "rhetorical code studies," as theorized by Mark C. Marino and Kevin Brock, respectively, have shown that the close analysis of source code can expand our understanding of the cultural implications of software beyond what interfaces reveal.[27] Additionally, as I have shown elsewhere, it is also possible to explore the gap between interface and algorithm by studying source code archives through large-scale text analysis.[28] Unfortunately, source code is not publicly accessible for most software, both today and historically. In its absence, examining the interplay between a technology's rhetorics and models of usability can bring to light the range of strategies and plans that are addressed through a concern for usableness and usefulness.

In addition to studying specific pieces of hardware and software alongside the way that their designers describe their functions, analyzing discussions by journalists and other users of their experience with personal computers can help highlight how rhetorics and models of usability work together to structure our relationship to digital culture. For example, Kirschenbaum's more recent work on the "literary history" of word processing is less focused on analyzing specific word processing software applications and more on understanding how authors in the early days of the personal computer structured their writing practices around particular pieces of hardware and software. As he notes in his preface, his goal is not to participate in the "easy or self-fulfilling narratives of technological progress" common among popular histories of computing; rather, in studying the contexts in which word processors were used and the people who used them, he is able "to offer an account of what was perceived to be at stake with word processing, and address the question of why the technology—which at first may seem little more than a welcome upgrade of the typewriter—proved so contentious."[29] Kirschenbaum enacts Latour's call to study technology in a state of crisis by studying literary authors who were using word processors before they were commonplace tools and highlighting how they responded to claims made by manufacturers and continually negotiated the role of computing in their writing. While Kirschenbaum's work is instructive, Deborah Brandt reminds us that the literary authors he discusses represent only a small subset of professional writers.[30] Thus, the kinds of questions that Kirschenbaum asks might be answered differently if we expand our view to include other contexts of early computer use. As I show, many concepts influencing our acceptance of transparency as user-friendliness are the product of a discourse that spans multiple contexts and that flows back and forth across highly technical circles, corporate offices, curriculum development, and domestic spaces. At various points, each of these contexts came to dominate conversations on usability, serving as a model for user-friendliness in the others.

Because more and more aspects of our public and private lives are being subsumed within digital culture, questions regarding how software systems shape our understanding of computation and its effects are not limited to digital media studies. My argument thus also builds on research in communication, library science, digital rhetoric, and writing studies by scholars like Siva Vaidhyanathan, Safiya Noble, Stuart Selber, Cynthia Selfe, and Annette Vee.[31] Their work may not engage as directly with transparent design or personal computing as my own; however, the questions they raise and the methods they employ help to illuminate what is at stake in pushing back against the politics of user-friendliness. Our ability to identify and articulate the nature of problems in our increasingly sprawling information infrastructure is in many ways constrained by the limitations imposed on our computer literacy by transparent interfaces; user-friendly interfaces have become our primary source for understanding how these systems work, and so developers can exercise significant control over what we know. Many of Big Tech's scandals are the result of a company's strategic misrepresentation of the large systems it manages. When an application, platform, or operating system is revealed to function differently from the way it is represented by its interface, we run up against the limits of our literacy. We struggle to identify, describe, or respond to those mechanisms that designers decide not to show us. Moreover, even when we do manage to uncover a mechanism that conflicts with the public image of the way a particular piece of software is supposed to function, designers can exploit the constraints on our literacy to deflect criticism of their technology. For instance, Google's engineers, as Noble has documented, appear to intervene invisibly in the company's search service so that the results it returns better align with a rhetoric of usability that represents its algorithms as objective and neutral.[32] So long as developers refuse to expose these systems to us, we are vulnerable to a form of digital gaslighting.

One key assumption driving the archival research I undertook for this book is that it takes considerable effort to build and maintain shared understandings of usability. Many early proponents of transparent design appealed to scientific principles to explain how to make use behaviors feel "intuitive," yet the feeling that use behaviors come naturally is the result of continuous efforts to align users' various understandings of computation with intended rhetorics and models of usability. As Emerson observes, transparent interfaces function like a "magician's cape, continually revealing . . . through concealing and concealing as it reveals."[33] Feature updates, redesigns, and new device models are always presented to us as somehow more friendly and more transparent than what we currently use. Each new feature that Apple, Google, or Microsoft introduces is presented as intended to enhance or improve activities we are already performing. Never do the accompanying announcements, documentation, or presentations suggest that we will have to restructure our present way of doing things to account for this next innovation even though, invariably, we do so each time. Even if the companies acknowledge that certain updates are indeed disruptive, they encourage us to treat them as momentary inconveniences and to quickly put them out of our minds so that we can minimize the amount of time we have to spend thinking about computing and focus on more important things. Perhaps the biggest challenge in recent memory to the maintenance of user-friendliness has been the COVID-19 pandemic, as it cannot be dismissed as a mere nuisance. Many people in the context of work and school from home have begun to realize, it has been reported, that wholly structuring our lives around these easy-to-use technologies is not as simple to do as their designers would have us believe.[34]

Each chapter of this book enacts some aspect of Susan Leigh Star's "ethnography of infrastructure" by surfacing the invisible work of personal computers and the continual interventions on our user behavior that developers engage in through them. I adopt Star's strategy of identifying and challenging the master narratives that position transparency as a universally applicable approach to user-friendliness. Star notes that master narratives can take many forms but that in practice they act as a "single voice that does not problematize diversity," a voice that "speaks unconsciously from the presumed center of things." This voice is present both in rhetorics and models of usability: in details of the technology itself, in records of activity surrounding or taking place across technology, and in representations of technology that are presented as if they are "literal transcript[s] about the process and progress of science."[35] Models of transparent design often serve as proof of the promises made in rhetorics of user-friendliness, which in turn drive the pursuit of new, more innovative models. The single voice that Star describes emerges from the way rhetorics and models of usability complement and sustain one another. Big Tech leverages this interplay to assure us that its genius and benevolent intent enable it to solve almost any cultural, social, or political problem through a creative application of technology.

Imagining an Unfriendly Future

A primary, though largely unspoken, assumption underlying the association between user-friendliness and transparent design is that computation and culture are distinct domains inherently in conflict with one another. The idea that computation and culture are utterly discrete domains is a powerful fiction that parallels the modern presumption of a dichotomy between nature and science identified by Latour.[36] Elite technologists leverage this idea of separate domains to justify their power by implying that only they can design hardware and software capable of resolving the tension between the two. Earlier characterizations of personal computing, by contrast, suggested that computation and culture were only made separate through efforts to control access to the former.[37] The term "personal computer" itself reflects a belief that there is a tension between the two domains and that users are being invited to participate in its resolution. Although the term originated in marketing copy for programmable calculators, by the mid-1980s it had supplanted the more technical-sounding "microcomputer" in advertisements for new desktop-sized machines.[38] Even IBM, a company portrayed by personal computing enthusiasts as trying to reshape the world to meet the demands of its computers, enthusiastically embraced this idea in its early 1980s marketing campaigns. However, we need to recognize that in practice there is very little that is "personal" about our computers. We install software written by someone else and have little agency to intervene in its operation apart from the fewer and fewer options afforded to us in increasingly vague configuration menus. One goal of this book is to push back against the idea that our computers are personal at all. In fact, one could argue that today's always online model of use has radically recentralized computing. The data we store on our computers may be unique to us, but the constant flow of updates ensures that the software on our devices resembles the ideal forms stored on a developer's master server. The rise of cloud computing, too, is transforming our devices back into terminals, windows into computational processes that are performed elsewhere.

Commercial developers were not the only ones to advance the idea of a conflict between computation and culture. Similar claims can also be found in early academic literature on usability in computing. Foundational writing in human-computer interaction often argued that efforts to make interfaces more "efficient" did little to change the fact that nonspecialist users found computers confusing and intimidating. Throughout the 1970s, computing professionals were trained primarily to develop software for corporate computing systems, but by the mid-1980s usability researchers began to consider the question of how to design computers for the general public in earnest.[39] As a discipline, human-computer interaction was founded on the recognition that software designers needed to account for a wider diversity of interests, backgrounds, attitudes, experiences, and identities than they had considered previously. This push continues today, as many within the field of human-computer interaction press for more "humanistic" approaches to design.[40] I argue, however, that designers historically have adopted rhetoric from the humanities in ways that trivialize rather than wrestle with the complex sociocultural concerns surrounding computing. I show how usability experts and designers often draw on humanistic concepts to raise concerns about technology but then propose alternative approaches that leave the problematic norms they are challenging in place, described now in different, more seemingly culturally engaged terms. While it is heartening to see technologists take up questions about the broader implications of their work, I worry that in practice these efforts merely draw on the prestige of the humanities to make yet another simplifying innovation seem more competitive in an entrepreneurial culture that increasingly claims to value interdisciplinary perspectives.

The idea that computation and culture are separate domains has also been invoked to argue that engagements with overly technical aspects of computer systems are of no value to users. In the user-friendly world we live in, we are encouraged to leave these aspects of computing to the professionals. By returning to the first decade of personal computing, I seek to understand the consequences of a digital culture that is founded on an intentionally and increasingly narrowly defined model of computer literacy. The long-term success of transparent design has been realized through a succession of models of usability that require users to know less and less about the mechanisms structuring their personal computing devices, offering them only as much technical language as necessary to make them usable and useful and thereby making it difficult to identify, let alone voice, concerns about the broader consequences of the information systems we interact with daily. Although Vee argues that we have entered a cultural moment in which programming knowledge has become a key part of our understanding of computer literacy, research by Selfe and Selber shows that computer education and training programs in the United States have since the 1990s largely focused on the use of particular applications for specific commercial and industrial purposes.[41] These conflicting definitions of computer literacy help to illustrate that computation and culture cannot be understood separately. No matter what an interface promises, our relationship to technological systems is necessarily complicated. Further, Antonio Byrd has shown that competing definitions of computer literacy can be leveraged to reinforce existing social inequities.[42] We must therefore begin to reflect on the political implications of seemingly unobjectionable design principles. As I argue, our ability to consider the interplay between the cultural, social, and technical aspects of computing is circumscribed by rhetorics and models of user-friendliness, which push us toward and away from specific lines of inquiry. We need to ask ourselves what it means that our foundational understandings of computation have been supplied to us by Apple, Google, and Microsoft, and how our dependence on them for even the most basic understandings of computing culture has influenced our critical study of digital media. Expanding access to structures and operations deemed to be "merely" technical is critical to our ability to confront the politics of personal computing.

Similarly, it is vitally important that we ask ourselves what we mean when we say that "personal computers should be easy to use." Rhetorics of user-friendliness are instrumental in allowing designers to "use" our computers to generate profit and structure our behavior past the point of sale. They use our personal computers to collect data, serve us advertisements, or gently coax us into making continual purchases. It is in their best interest for these devices to be seen as friendly, which is to say nonthreatening and largely free of frustration, so that we will invest more of ourselves in them. By suggesting that complex social, cultural, and political problems associated with computation can be solved through user-friendly design, Big Tech has been able to impose an ethic of expediency onto American understandings of digital culture that displaces all other approaches, which are deemed a threat to our progress toward a techno-utopia. That computers appear to get easier or more responsive each year is treated as proof of the necessity of Big Tech's methods. While I am not advocating for a return to a time when only a handful of people could access computers, there must be a way to balance our concern for the potential for abuse of transparency by developers with the real need to ensure that computers—which are key to most forms of social and cultural advancement—remain widely usable and useful. In other words, we must consider the rhetorical history we invoke implicitly whenever we say that computers should be "easy to use" or need to "just work." In the pages ahead, I insist that we open ourselves to imagining an "unfriendly future," a form of computing that values complex engagements with technology and that actively supports the agency of users and communities to critique and intervene in computation. To do so, we must first learn to recognize the way that user-friendly design simultaneously appears to address and to conceal responses to a wide range of concerns beyond usability.

The Structure of This Book

Each chapter in this book discusses a specific moment in the rhetorical history of user-friendliness, with a focus on the introduction of new rhetorics and models of usability that contributed to the consensus around transparent design. In chapter 1, I introduce my archive and explain how I understand the relationship between rhetorics and models of usability as well as examine how user-friendliness initially appears in popular American computing magazines as a technical problem-solving strategy in the context of system design. According to its earliest proponents, transparent design produces a loss of information that "comes for free," introducing added utility to a system by automating and concealing processes that were deemed to be merely technical inconveniences. As I argue, this approach was eventually proposed as a way to address complex concerns about computer literacy. Transparent design would function as a kind of hack, obviating the need to explain how an application or a system worked to users by constraining user agency such that the more limited set of skills and understandings available to them would be readily accepted as natural or intuitive. Importantly, this chapter also begins to disentangle the rhetorical history of user-friendliness from the technological development of GUIs. I argue that this separation is crucial because we must recognize that user-friendliness implies more than simply making computers "easy to use." Often, it becomes a way of persuading users to accept certain power relationships with software developers as a condition of use. I conclude by considering how today's always online models of usability afford designers the ability to invisibly normalize user behavior in ways not anticipated in early discussions of user-friendliness and transparent design.

My tracing of the social construction of transparent design begins in chapter 2 with a reexamination of the rhetorical origins of the personal computer revolution. Popular histories of hobbyist computing in the mid-1970s, like Steven Levy's *Hackers: Heroes of the Computer Revolution*, helped construct a narrative that Silicon Valley draws on today to justify its authority through appeals to personalness and universal usability. Levy's narrative in particular imposes a monolithic political framework onto early personal computing enthusiasts, who he argues were inspired by the writings by Stewart Brand, Ted Nelson, and Ivan Illich. I challenge that framework first by highlighting points of contention between the three and then by turning to an analysis of hobbyist writing in the *People's Computer Company*, the *Homebrew Computer Club Newsletter*, the *Mark-8 Users Group Newsletter*, and *Computer Notes* in order to consider the different rhetorics of usability they developed to explain their social and technical practices. Each engages with Brand's, Nelson's, and Illich's ideas in distinct ways, and they respond differently when faced with technical issues that challenge their expressed values. As I show, personalness and universal usability begin to emerge as part of an ethic of expediency. When presented with an expedient solution, many hobbyists opted to revise their rhetorics of usability to justify an apparent contradiction between their goal of decentralized computing and the convenience of centrally managed components. They thereby ended up linking personalness and universal usability in support of a recentralization of computing that is first visible at the West Coast Computer Faire and in Apple's early advertisements.

Outside of countercultural circles, the personal computer revolution was also a popular point of discussion; however, mainstream American print media often described the advent of personal computing in quite different terms. As I show in chapter 3, many journalists and hardware manufacturers in the late 1970s and early 1980s offered a much more conservative vision that framed personal computers as a tool for business or convenience in the home. Early manufacturers of "appliance computers" like Tandy, Commodore, Atari, and Texas Instruments developed rhetorics of usability that promised "friendlier" computers that would be immediately usable by and useful to novices. Through a review of interviews, advertisements, product reviews, catalogs, and other print texts that circulated around early appliance computers, I show how these companies relied on the idea of ease of use to suggest that models of usability needed to be structured around proprietary control to ensure increased access to technology. Even though computers like the Tandy TRS-80, Commodore PET, Atari 800, and Texas Instruments TI-99/4A did not have a lasting technological influence on personal computing, they did help to establish a strong association between models of usability that intentionally limited user agency and the rhetoric of user-friendliness.

In chapter 4, I examine the rhetorics and models of usability associated with the IBM PC and Apple Macintosh in the context of what some journalists described as a "computer literacy crisis." Many new users were finding that the experience of using their personal computers did not live up to the promises of the late 1970s and early 1980s. While many companies presented their products as "friendly" in response to these concerns, IBM and Apple in particular positioned their computers as answers to this literacy crisis. Following a hurried design process, IBM leveraged its robust documentation to develop a rhetoric of usability that suggested the machine was designed to allow users to establish a personal relationship with their computers through a "structured program" that would help them develop their computer literacy. The Apple Macintosh was portrayed more simply as offering a model of usability that would make computer literacy a thing of the past. Rather than promise users that the Macintosh would help them learn about computing, the rhetoric of usability that Apple developed suggested that computation and culture were in conflict. The tension could only be resolved by a more user-friendly approach to design that did not require users to relearn familiar concepts in computational terms. In examining these two machines, I show how the rhetorics of usability associated with each ultimately led journalists to interpret problematic aspects of their designs as desirable, which in turn illustrates how popular understandings of what it means for computers to be usable and useful are shaped by the complex interplay between rhetorics and models of usability.

Were it simply one company defining user-friendliness according to the principles of transparent design, we might live in a world where we had more choices in interface styles. Yet as I show in chapter 5, an important reason why our popular understanding of user-friendliness has remained so durable since the 1980s is that academic researchers in the field of human-computer interaction developed principles of usability that naturalized transparent design. The field of human-computer interaction emerges as a series of efforts to critique approaches to usability that were influenced by industrial engineering, artificial intelligence, and cognitive psychology. Many foundational theorists in human-computer interaction—Ben Shneiderman, Donald Norman, Terry Winograd, Fernando Flores—developed rhetorics of usability that positioned their work as better suited to address the increasingly diverse needs and interests of novice computer users. As I argue, however, a number of the problematic assumptions found in earlier usability research ground the principles of usability that these foundational writers propose, with the result that the very concerns about user identity and contexts of use that they claim to address are "screened out." Foundational literature in human-computer interaction thus co-opts language from the humanities to naturalize transparent design. This problematic use of humanistic rhetoric is even visible in work that promises to apply the lens of critical theory to usability. As I argue, Brenda Laurel's work integrating theories of drama and usability reproduces the same assumption that interfaces should passively constrain user agency in the interest of ensuring a singular experience. By contrast, Lucy Suchman's research approaching usability from the perspective of ethnomethodology challenges efforts by designers to maintain a boundary between themselves and users. Rather than see design as a way to solve problems, Suchman instead frames design as an ongoing process of asymmetrical communication that continues even after a technology is built. She insists that designers should look for ways to foster continuous discourse with users rather than assume that they are uniquely positioned to determine what is in their best interests.

I then conclude by considering what an alternative, "unfriendly" future might look like in light of the issues raised in previous chapters. To realize this future, I argue that we must recognize how user-friendliness conceals the consequences of design. If we approach usability with the goal of valuing complexity rather than simplicity, we can work toward realizing a form of computing that privileges political transparency rather than transparent design.

1

On the Origins of User-Friendliness

User-friendly design can be understood broadly as a set of strategies intended to simplify software by structuring interfaces to complement users' existing understanding of specific situations, activities, or tasks. A "friendly" interface is said to be tailored to users' needs rather than requiring users to acclimate themselves to the system's complexities. The goal of this approach to design is to ensure that users immediately recognize a technology as usable and useful.[1] Ideally, use behaviors feel "natural," intuitively aligning with the way users have always approached the tasks that are now represented on screen. In more specific terms, this feeling of naturalness or intuitiveness occurs when we experience hardware or software as "transparent," disappearing from our awareness during use such that we do not understand ourselves to be operating a computer nor pause to reflect on how the computer is continually guiding our actions and shaping our thinking.[2] Critics and historians of digital media often associate transparency in personal computing with the emergence of graphic-user interfaces (GUIs,); however, early discussions of transparent design precede not only GUIs but also personal computing itself. By tracing the development of the rhetoric of user-friendliness across engineering literature and early personal computing magazines, I show in this chapter how user-friendliness came to be strongly associated with transparent design as programmers in the late 1970s and early 1980s began to apply problem-solving strategies that had proven successful with narrowly defined technical questions to comparatively abstract sociocultural concerns about usability.

Importantly, most conventional understandings of user-friendliness suggest that transparency can only be achieved through interfaces that provide simplified representations of a computer system. Since the 1980s, one of the most vocal advocates of this approach to design has been cognitive engineer Donald Norman.[3] He has argued throughout his career that most user errors result from interfaces that do not signal their specific, intended use behaviors to naïve observers. Norman notes that computers present a unique challenge for design in that there is no "hard" link between their mechanisms and their interfaces, and he sees this challenge as an opportunity for a new approach to usability. Because "the computer is unusual among machines in that its shape, form, and appearance are not fixed," he explains, its interface can take on any appearance "the designer wishes." The computer, he continues, "can be like a chameleon, changing shape and outward appearance to match the situation. The operations of the computer can be soft, being done in appearance rather than substance."[4] Software designers, in other words, should not feel the need to design interfaces that faithfully represent a system's internal functions. On the contrary, doing so is often counterproductive if our goal is to reduce user error given that the relationship between those basic mechanisms and the application's intended use behaviors is not always immediately clear.

Although few are as blunt as Norman, his work is representative of the prevailing assumption that an understanding of how computers work is of limited cultural value compared to the ability to get things done with them. This view of computing has two important implications. First, simple engagements with computing are always understood to be more valuable than complex ones because complex ones diminish our sense of a technology's utility. Second, the skills and understandings necessary to navigate the interface can be wholly divorced from those necessary to understand the algorithmic systems that support it. These two implications have served as guiding principles for commercial software design since the mid-1980s, and exploiting the gap between algorithm and interface that Norman points to has become a key strategy for realizing them in application interfaces. However, it is important to recognize that by exploiting this gap, designers are not merely building applications but also curating digital culture, ostensibly for the benefit of everyone else.

Our understanding of digital culture is ultimately structured through the rhetorics of usability provided to us by hardware and software designers. The language embedded in and circulating around interfaces informs our everyday understandings of what computers can and cannot do, of the scale and scope of our digital agency, and of what roles computers should or should not play in our lives. I use the term "rhetoric of usability" to refer to the language that is distinct from the digital and material affordances of a technology but that nonetheless influences our use practices as much as if not more than those affordances. When discussing the way that the digital and material affordances of a technology encourage or restrict specific sets of use behaviors, I use the term "model of usability." In developing two distinct terms, I do not mean to imply that either is wholly separate from the other. As I show, rhetorics and models of usability each inform our interpretation of the other. Rhetorics of usability function as a kind of "top layer" in a technology stack, conceptually located between our internalized understanding of a technology and the interfaces presented to us. We interpret models of usability by drawing on the discursive concepts that circulate around them; however, we also evaluate rhetorics of usability based on our embodied experience of actual technologies, accepting or rejecting those rhetorics depending on how well they seem to explain those experiences. Importantly, rhetorics of usability can further complicate problems of understanding produced by the gap between screen and code. While they can be leveraged to normalize user behavior by pushing us to accept features as natural, necessary, or even desirable that might otherwise be received as anticonsumer, antidemocratic, or otherwise disruptive, they can also help to grow and organize resistance to technocratic power. We therefore must recognize that rhetorics of usability promising user-friendliness always advance a broader array of concerns than simply "ease of use."

In this chapter, I explore how a logic of simplification used to reframe problems in system design came to be mapped onto concerns about usability. Through a rhetorical analysis of engineering literature and early computing magazines, I show how designers moved from explaining how to make components transparent to the system to proposing that the system could be made transparent to the user. Initially, transparency was only invoked to explain how to use abstraction and encapsulation techniques to develop simplified representations of components that would minimize the cognitive labor of managing hardware states while programming. If the simplified representation was outwardly functionally similar to the original, then simplification was understood to "come for free" in the sense that it did not require programmers to further reconceptualize their designs or modify the functionality of other parts of the system. This same logic was eventually applied to user interfaces under a similar assumption that a simplified interface would not require users to adjust or change the cultural or social activities they were engaging in to accommodate the overly technical aspects of software. By the mid-1980s, user-friendliness became a popular marketing term that associated the principles of transparency with a vision of socially and culturally responsible design. This vision persists today as part of a Silicon Valley, start-up mythos that assumes that the integration of computation into our daily lives is always beneficial: each new innovation cuts through some lingering complexity in our lives, offering a simpler approach that can be managed on our behalf by some new invisible automation.

Another subject I consider in this chapter is the role that rhetorics and models of usability play in maintaining the perception that technical reasoning is politically neutral. The idea of neutrality here is nuanced. We do not treat technology as having no politics; rather, it is more common in US culture for technology—and especially digital technologies—to be portrayed as inherently beneficial to all sectors of American society. Within this framing, personal computing technologies are presented as tools that do not play ideological favorites: anyone who chooses to use them will find their lives improved and everyone is welcome—and encouraged—to use them. This form of supposed neutrality is supported by the interplay of rhetorics and models of usability associated with user-friendliness and transparent design, which obfuscates the costs and consequences of design. Like Steve Woolgar, I understand the production of technology to be a process through which social relations are "frozen" into a material / digital form. It is important to recognize that even as many usability experts do recognize and try to engage critically with the social and cultural dimensions of their work, they are but one source of influence in the corporate development norms that dominate the design of personal computing technologies. In practice, software development is a process of negotiation during which design teams "configure" users, defining their intended use behaviors by balancing some sense of users' needs against their own economic and political interests. As Woolgar notes, technologies are texts in that they are open to interpretation by users, but he cautions us to recognize, too, that software development teams work hard to ensure that users will be more likely to accept a "preferred" reading.[5] Designers, particularly those working in commercial contexts, leverage both the discourse surrounding computing and the material / digital affordances of the software they build to push users to align their understanding of a given technology and their use behaviors in support of larger purposes that extend beyond their immediate awareness, whether that be reaching a quarterly business goal, implementing a long-term competitive maneuver, or attempting to adjust a company's public image. So long as the software appears to behave as promised, users need not concern themselves with how it functions, what sort of data their use behaviors are intended to generate, nor what other purposes it may be accomplishing beyond those tasks represented on their screens. Apart from those rare moments when the veil of transparency is lifted—often inadvertently—we typically cannot see for ourselves how our software functions. This information asymmetry affords software developers tremendous cultural and social power over the contexts that become dependent on the applications, platforms, and frameworks they build.

This chapter begins by revisiting the origins of user-friendliness. Whereas many studies of transparent design in the humanities focus on its relationship to visual art and GUIs, I highlight how its core principles were adapted from programming techniques that were already being practiced before the rise of highly visual interfaces. I also provide a brief overview of how the editorial policies of early American computing magazines during the late 1970s and early 1980s were revised in response to a growing interest in personal computing by individuals who did not come from technical backgrounds. The changes correspond to a sharp increase in descriptions of hardware and software as "friendly". Ultimately, a consensus formed that transparency was the best way to realize user-friendliness as more and more technologists began to adapt to the complex problem of computer literacy a problem-solving strategy that had proven successful in simplifying system design tasks. As I conclude, one important unacknowledged consequence of the consensus that user-friendliness should be realized through transparent design has been a narrowing of the public's computer literacy in ways that has helped to naturalize the rhetorics and models of usability associated with user-friendliness.

Transparency's Complex Origins

A first step to challenging user-friendliness as transparent design is to recognize that simplification does not necessarily require graphical objects like icons. In 1983, a developer named Sam Edwards who worked for Software Publishing Corporation, a company known for its popular text-based productivity applications, argued that the main requirement of a good interface is that it "stays out of your way by not drawing attention to itself." Referring primarily to decisions he made while designing pfs: Write, he notes that his goal was to offer users only what they "need to know": "The fewer the choices on the menus and prompt lines, the less you'll have to think about before making your choices. The less a program does, the fewer things can go wrong with it."[6] Additionally, while Edwards never uses the term "transparency," he describes well-designed interfaces as ones that allow users to "concentrate on [their] work and not on using the program" by limiting access to information, producing user experiences that are "natural and unexceptional."[7] Edwards does note in his conclusion that new developments in high-resolution graphics will likely help with these efforts; however, his article shows how many of the same ideas we now associate primarily with user-friendly GUIs were being practiced in software developed for earlier command-line-based environments. Although transparent design's principles are often assumed to have emerged as part of the development of early GUIs, these ideas have a separate history that precedes machines and software developed by companies like Xerox, Apple, VisiCorp, and Microsoft.

The association of user-friendliness and transparent design with GUIs also supports the master narratives promoted by developers of early graphically oriented computer systems like Xerox and Apple. In the early 1980s, Xerox's Star and Apple's Lisa and Macintosh were touted by their designers as technologies that would finally allow computers to be accepted outside of highly technical, hobbyist circles as a cultural tool. The idea that an understanding of computers as cultural tools depends on the visuality of computing has been widely accepted in American culture. A significant body of digital media scholarship is even dedicated to understanding GUIs within the context of the aesthetic principles of Western visual culture.[8] Jay David Bolter and Rachel Gromala argue, for example, that a "desire for transparency" can be found in discussions of perspective among European painters as early as the Renaissance and that "computer graphics and interface design continue this history."[9] They go on to assert that prior to foundational work in GUI technologies by Douglas Engelbart and Alan Kay, "computer applications did not have consciously designed interfaces at all" in the sense that software was "not designed to provide the user with a consistent experience."[10] However, assertions like this one implicitly dismiss decades of research in human factors engineering that sought to consciously design interfaces that would consistently improve operator performance by reducing physical and cognitive strain. Such assertions also illustrate the powerful influence of claims made by both Xerox and Apple as they each worked to bring to market the first commercial GUIs. For example, Steve Jobs's biographers have written at length about how he wanted the Macintosh's visual aesthetic to be recognized as a work of art, as something that belonged not just on desktops but in museums.[11] Xerox engineers describe their work on the Star interface in 1982 as the first to develop "the fundamental conceptions (the user's conceptual model) *before* software is written rather than tacking on a user interface *afterward*."[12] Similarly, Apple would boast in its brochures for the Macintosh that it represented "the first time in recorded computer history" that a design team was able to "teach computers about people, instead of teaching people about computers" by designing both the machine's hardware and software systems around a desired interface experience.[13] As Mar Hicks has observed, the reliance on corporate sources for our understanding of American computational history has made it difficult to locate material that doesn't privilege a company's preferred narratives.[14] Dissociating transparency from GUIs can thus better position us to avoid inadvertently promoting the same narratives that software designers advance within the rhetorics of usability that they leverage to maintain their positions of power over digital culture.

It is important to recognize, too, that the history of GUIs includes interfaces that were not transparent. Although digital media studies often locates the origins of transparency in Vannevar Bush's speculative descriptions of the Memex, Elizabeth R. Petrick notes that Bush's Memex represented a different design paradigm: user augmentation rather than user-friendliness, which was "the opposite of an invisible interface; it is literally at the user's fingertips, in front of them, in a way that cannot fade into the background."[15] Digital media scholars often position Engelbart's writings about his oN-Line System (NLS) as a bridge between Bush and the first descriptions of modern GUIs. Yet Lori Emerson's discussion of Engelbart's software suggests that his NLS would not be recognized today as user friendly. Rather than pare down functionality, the NLS added a graphics layer for the purpose of providing users with increased options to control how much information was provided to them at any given time and to determine more directly the form that information took.[16] Histories of personal computing celebrate Engelbart's demonstration of the NLS; however, it is important to recognize that his software saw limited use outside of his laboratory. The NLS's only known users worked in the US government's Advanced Research Projects Agency, and many of them complained that it privileged "expert users." They suggested, too, that the less powerful applications they had used previously performed the same tasks as the NLS but "much faster and with more flexibility."[17]

While this work is helpful in understanding how our definition of what counts as an "empowering" interface has changed throughout the twentieth century—especially in the sense that it has not always implied "ease of use"—it is important to recognize that the influence of Bush and Engelbart on popular understandings is remote. They are only briefly cited as inspirations on usability theory, and the specific principles they outline in their writings are rarely engaged with directly by researchers like Norman. Their work is also largely absent from the writings of those hobbyists who are depicted as the source of the personal computer revolution. If we are to approach the consensus around user-friendliness as one that was largely derived from the specific hardware and software that were available to consumers, then we need to look past the speculative and experimental systems associated with early visionaries and instead turn our attention to the specific rhetorics and models of usability that circulated within popular culture. There, we can begin to see that during the first decade of personal computing the ability of the average person to use a personal computer was to many software developers and industry observers just another problem that could be solved with the right technology, another problem solvable by a clever hack. Within tech culture, hacking plays an important role in the rhetorics of usability that Silicon Valley elites circulate to portray themselves as uniquely suited to solve today's complex problems. The term itself generally refers to problem-solving methods that radically reformulate a complex problem so it lends itself to a comparatively simple, technologically oriented solution.[18] Transparent design was first proposed as a kind of hack: a way to simplify programming problems by strategically misrepresenting components or hiding processes so that software developers would not have to manage them directly. This same problem-solving strategy was then mapped onto the complex problem of teaching the general public how to use computers. Rather than develop interfaces that communicated computational concepts in new ways, designers instead opted to hide computing.

"Transparency" in Engineering Literature

Founded in 1947, the Association for Computing Machinery (ACM) is an international professional society for computer scientists and engineers that is largely made up of academic researchers and students but that also includes members who work in government and industry. Among its publications are conference proceedings, academic journals dedicated to particular subfields of computing, and magazines of general interest to computing professionals across a variety of fields and industries such as *Communications of the ACM*. As an archive, the ACM Digital Library database offers both a cross-sectional look at how specific concerns are addressed across various subfields within computer science and a sense of how consensuses form around those concerns over time. Within the database, references to "transparency" or "transparent design" appear in several different contexts. The terms are commonly used without being defined, suggesting that the concepts associated with transparency were familiar to readers of the ACM's publications. Although there are some subtle variations in its usage, in practice transparency in these early examples refers to the idea of introducing a new feature to an existing system in a functionally invisible way.

References to transparency can be found both in articles discussing established computing principles and proposals for new techniques alike. A short article published in 1965 in *Communications of the ACM*, for example, describes how transparent modes operate in terminal network hardware. Here, "transparent mode" refers to a method that many networked terminals use to incorporate a secondary control schema into their character encoding systems. Under normal operation, the terminal recognizes data that should be decoded for display by looking for "control codes," or characters that designate the boundaries of a data structure in a serial transmission. Introducing a secondary schema, however, allows control code characters to be treated as normal data and so displayed to users for diagnostic purposes. Apart from using a different set of codes to enter and exit from the secondary schema, the packet encoding system functions as normal, meaning that other hardware and software components involved in sending or receiving data do not need to be altered to support this feature.[19] While transparent mode communication was a standard feature of network hardware, there are also references in ACM publications to transparency in descriptions of other aspects of system design. A paper from a 1977 conference proceeding, for instance, discusses software monitoring tools that would be "completely transparent to almost all target programs." These tools would rest between existing components, analyzing the data that is passed between them but not altering it. This design would allow the monitoring software to invisibly observe the efficiency of a system without disrupting its normal function.[20] While the contexts are different, in both cases software developers suggest that a functional invisibility allows for interventions that preserve the original intent of the system's designers.

In addition to providing descriptions of components that were transparent to the system, computing engineering literature also refers to design techniques that would make the system transparent to the user. Because prior to the 1980s most if not all computer users were assumed to be programmers, many of the references to user-transparency in the ACM publications pertain to methods for simplifying the management of system resources during software development. As early as 1974, for example, engineers proposed automating certain routine processes that were "small nuisances." Making them "user transparent" would "relieve the user" of having to manage routine configuration tasks that could be performed automatically at system start-up.[21] Initially, discussions of user-transparency were directed mainly toward small tasks or limited contexts. By the mid-1980s, however, transparency was being invoked as a general theory of system design. A 1985 essay describing the design of a distributed operating system, for instance, notes that transparency is a "key concept" for usability: "a distributed system is one that looks to users like an ordinary centralized operating system but runs on multiple, independent central processing units. . . . In other words, the use of multiple processors should be invisible (transparent) to the user."[22] Strategically concealing components from programmers and systems administrators would reduce their cognitive burden while still allowing them to fulfill the essential functions of their jobs. This technique would similarly obviate the need for either type of user to receive additional, specialized training to operate multiprocessor systems. Nothing would be lost by not requiring these users to manage algorithms across multiple processors. Making complexity invisible through a combination of automation and simplified representations would thus allow users to realize more readily their intended purpose when using a system.

"Transparency" in Personal Computing Magazines

References to transparency can also be found throughout early American computing magazines. Initially, these publications were devoted almost exclusively to detailing hobby projects for personal computer kits. By the 1980s, however, they had turned their attention to commenting on the people, companies, and products participating in the nascent personal computer industry. While some American electronics and radio magazines like *Popular Electronics* and *Radio Electronics* discussed circuit diagrams and provided information on how to locate instructions or parts to build home computers, the manufacture of complete kits after 1975 by companies like Micro Instrumentation and Telemetry Systems; Processor Technology; and Information Management Associates, Inc. led to the publication of specialized computing magazines like *BYTE* and *Creative Computing*. Most of the projects they covered were intended for machines built from Intel 8080 processors; however, they also provided software in the form of BASIC source code that could be used with both home computers and time-shared systems if readers were able to adapt it to the idiosyncrasies of the specific interpreter they were accessing. Other magazines that began publishing after 1977 tended to explicitly favor other hardware standards. For example, *80 Microcomputing* focused on the Tandy TRS-80, built around the Zilog Z80 processor, and *Compute!* focused on machines built around the MOS Technology 6502 processor like the Apple II and the Commodore PET.

Their different hardware focuses notwithstanding, most American computing magazines offered similar content. In this respect, *BYTE* magazine, which began in 1975, serves as a good representation of early US computing journalism both because it was one of the earliest and longest lasting and also because changes to its editorial policies align closely with readership trends occurring across most American computing magazines.[23] *BYTE*'s first editor, Carl Helmers, explained in a 1978 editorial that he wanted the magazine to "provide readers with a continuing stream of novel ideas and information about computers and related fields. The assumption made about the reader is that he or she possess curiosity combined with a willingness to experiment."[24] Initially, *BYTE* focused on supplying users with hardware how-to guides, tutorials, and sample software source code for readers to type up and to try out on their own computers. In addition to hobby-related content, a typical early issue of *BYTE* also included short news articles announcing how manufacturers were making new parts or services available and much longer opinion pieces in which contributors debated the merits of new programming languages, tried to persuade readers to consider some new algorithmic technique, or speculated about new applications of computing in business, scientific research, and education.

By the 1980s, however, fewer and fewer readers of popular computing magazines were hobbyists looking for help with home projects but were instead consumers seeking guidance in navigating a growing market for hardware and software and information about the latest business applications. In 1981, *BYTE*'s then editor in chief, Chris Morgan, explained that the magazine would be moving to include more regular review content "in response to reader surveys that show [an] increasing interest in the many new products flooding the market."[25] Although Morgan assured his readers that review content was not going to replace the type of article that had become the "mainstay" of *BYTE*, projects, guides, and tutorials had all but been formally phased out of its issues by that point. Morgan's successor, Lawrence J. Curran, would revisit the matter two years later, explaining that the magazine would be reporting more industry news because he understood the magazine's readers to include "professionals in fields such as law, medicine, accounting, and business management who rely on computers as personal tools in their work; scientists and engineers in the computer industry who regard computers as essential development aids on the job; and those who use personal computers for nonvocational pursuits." Curran pauses for a moment here, noting that this last group has often been referred to as "hobbyists, although that term is subject to careful reexamination today."[26] Most home users, he suggests, are no longer building their own machines or writing all their own software. By the 1980s, many novice users were drawn to home computing because manufacturers had begun to emphasize how off-the-shelf parts and "canned" software would allow them to forgo direct engagements with the overly technical aspects of the system. Although there is likely some overlap between readers of computer magazines and professional publications like those in the ACM's database, computer magazines by the 1980s were increasingly directed at readers without formal training or professional experience in computing. Reader surveys, for example, indicate that while at least a third of *BYTE*'s readers self-identified as engineers, computer technicians, or programmers, the majority came from a wider variety of professions.[27] Anecdotal accounts of the people who frequented computer shops in the early 1980s similarly suggest that an increasing percentage of people interested in computing did not have a formal background in computer science or engineering.[28]

Importantly, these magazines were also full of advertisements that reveal how user interests were represented within the growing industry. Most included as part of their backmatter a "reader service," which consisted of a complete index of advertisers (figure 1.1). Readers were instructed to fill in their contact information and to mark the boxes on the index next to the advertisers they wanted more information about and then clip out the index for mailing. The magazine would later forward the mailing addresses provided by readers to the indicated advertisers, who would in turn send catalogs, brochures, and other informational materials to readers directly. The instructions accompanying these readers services also indicate that they likely served as marketing research, providing data that could be used to attract new advertisers and to negotiate their rates. By the 1980s, American computing magazines often devoted as many pages to advertisements as they did to news, reviews, and discussion pieces. By this point, the indices in the back of many American computer magazines took up one or more entire multicolumn pages. Later iterations of reader services included postcards that were crammed from margin to margin with a list of numbers corresponding to an index entry that readers could circle and return. Index numbers could be found near the back of the magazine and were often also printed onto the advertisements themselves in a bottom corner.

These advertisements are a valuable resource for understanding the various rhetorics of usability that circulated around personal computers in the late 1970s and early 1980s, as they reveal a developing consensus around user-friendliness and transparent design. However, the sheer volume of them poses a challenge for analysis. In addition to manually reviewing early magazines for articles discussing transparency, I also performed a simple frequency count looking for instances of the term and simple variations. For these counts, I focused on a single publication, *BYTE,* in order to avoid complications introduced by differences in the typical number of pages in an issue, how many issues a year a magazine published, the quality of the scans

To get further information on the products advertised in this issue of BYTE merely tear, rip, or snip out this advertiser index, fill out the data at the bottom of the page, mark the appropriate boxes, and send the works to BYTE, Peterborough NH 03458. Readers get extra Brownie Points for sending for information since this encourages advertisers to keep using BYTE – which in turn brings you a bigger BYTE.

ADVERTISER INDEX

- ☐ ACM CII
- ☐ AP Products CIII
- ☐ Babylon 86
- ☐ Delta 43
- ☐ Godbout 8, 60, 61
- ☐ Hickok 48, 49
- ☐ James 42, 82
- ☐ Martin Research 1
- ☐ Meshna 93
- ☐ Micro Digital 2
- ☐ MITS CIV, 7, 71, 81
- ☐ Processor Technology 83
- ☐ RGS 59
- ☐ S.D. Sales 80
- ☐ Scelbi 38, 39
- ☐ Solid State 89
- ☐ Sphere 94, 95
- ☐ Suntronix 91
- ☐ Wahl 70

To help the editors with a profile of the readers — what type of work do you do?

Have you a microprocessor running yet? and which, if so?

Messages for the editor:

Reader's Service
BYTE
Green Publishing Inc.
Peterborough NH 03458

SEPTEMBER 1975

BYTE acquired via
☐ Subscription
☐ Newsstand
☐ Stolen

Please print or type.

Name ____________________

Address ____________________

City ____________ State ______ Zip ______

Coupon expires in 60 days . . .

Figure 1.1. The "reader service" mailer as it appeared in the September 1975 issue of *BYTE* magazine, its first.

of publications, and rates of character recognition failure across them. A frequency count of "transparent" and its variants shows that mentions of transparency can be found in *BYTE* dating back to its earliest issues (figure 1.2). The term became more common over time, with a significant increase in frequency in 1982. Much like the examples in the ACM library, transparency in *BYTE* is typically referred to without being explicitly defined. This is especially true in issues published before 1977, when the only computers available to consumers were assembled from kits. Closer examination of the results shows that initially "transparency" could refer to three different topics: transparent modes in terminal hardware, the functionally invisible design of system components, and shading techniques in computer graphics. After 1980, the term "transparency" and its variants increasingly show up in *BYTE* in discussions about the benefits of making specific mechanisms, components, or operations invisible and is commonly described as something that comes for free. Hiding those aspects of computing that are determined to be overly technical from users can only improve their experience. If automated properly, making them phenomenologically transparent—outside of users' awareness during operation—can also allow them to be transparent to human intention, as their invisibility will only improve users' ability to realize their goals when using a personal computer.

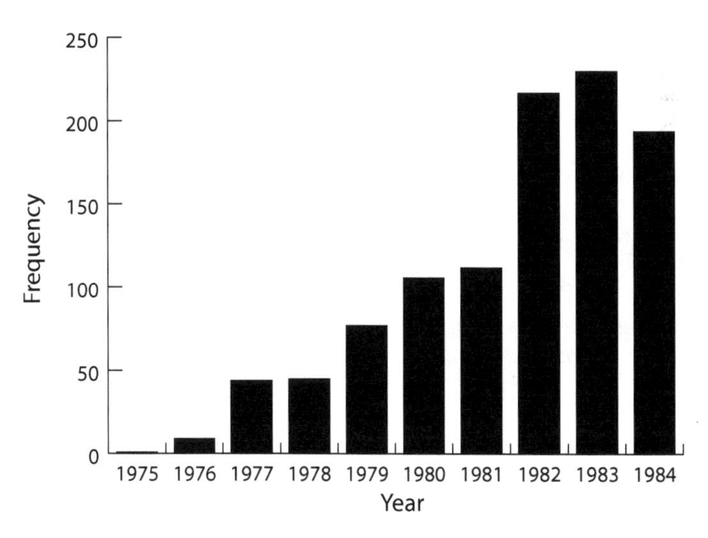

Figure 1.2. Annual word frequency counts of "transparent," "transparently," and "transparency" in issues of *BYTE* magazine, 1975–84.

The first mention of this kind appears in January 1976 in an advertisement for a computer shop announcing the mail-order availability of add-on chips for the Intersil 6100 (or IM6100) microprocessor. The brief copy promises simply that "transparent control panel chips" will allow the "IM6100 to free itself of control panel supervision."[29] In June of that same year, *BYTE* printed an entire article devoted to the IM6100 that explained the chip's design. Briefly, the IM6100 was designed to emulate the PDP-8 computer system, reproducing the functionality of the cabinet-sized "minicomputer" on a palm-sized microprocessor chip. Before the first home computer kits became available in early 1975, Digital Equipment Corporation's PDP-8 was a popular minicomputer that served as a focal point for experimentation by a significant number of hobbyist users groups and is commonly remembered as the first commercially successful minicomputer.

In addition to supporting the same machine code instructions as the PDP-8, the much smaller and cheaper IM6100 had additional programming features that supported the automation of PDP-8 subsystems. The article provides some basic instructions on how control panels, bootstrap loaders, terminal monitor programs, and other low-level system components could be "implemented [using the IM6100] in the same fashion, completely transparent to the normal PDP-8 mode of operation."[30] Like many early computer systems, the PDP-8's control panel was its primary interface, consisting of a series of lights and switches that could be toggled to represent the "op-codes" that issued instructions directly to the system's processor. Crucially, these panels also had to be manipulated to start a system and "bootstrap" into states wherein the computer could support higher level programming languages like BASIC. Early PDP-8 users described the start-up sequence as something that only the one or two most experienced people in a lab could manage.[31] The "transparent" chips mentioned in the ad would thus "free" the IM6100 of control panel supervision by automating boot sequences that would otherwise have to be performed by the user. The rest of the system could then function as normal; no modifications would be required to support the chip's automation features, and the system would perform no differently once booted than if a trained user had started it up. Eliminating the need for users to manage the complex start-up sequence thus increased their ability to take advantage of the IM6100's emulation of the PDP-8's higher-level functions.

Transparency is also frequently mentioned in accounts of the development of a simplified external representation of a component's internal function. By far, the majority of these mentions are found prior to 1980 in descriptions of random-access memory (RAM). Beginning in 1977, advertisements began to appear in *BYTE* for "DRAM" that promised a "transparent refresh, which means the memory looks static to the outside world."[32] In contrast to the advertisement for the add-on chips for the Intersil 6100, this one includes an explanation. Its very presence, despite its brevity, suggests that advertisers wanted a wider range of readers to recognize the benefits of the technology. In *BYTE*, the closest to an in-depth description of transparent refresh DRAM can be found in a 1977 article by Stephen Wozniak in which he details certain choices he made when designing the Apple II. Most digital computers, he notes, use a type of memory comprised of "capacitive storage elements" that "must be periodically recharged ('refreshed') to prevent the information from disappearing."[33] This type of memory is typically referred to as "dynamic" memory (or DRAM). Although "static" memory (or SRAM) was available and easier to program with, it was much more expensive and therefore much less common in microcomputers. Typically, he continues, the refresh cycle in DRAM is automated; however, because the individual storage elements are unavailable for reading or writing during refresh, programmers working in low-level languages (i.e., those in which programmers define and control hardware states directly) had to factor the refresh cycle's timing into their algorithms. Wozniak points out that it is possible to implement a "hidden refresh" by executing refresh cycles whenever a program scans through memory as part of an address look-up. When implemented in this way, "refreshing of the memories happens to come for free and is totally transparent to the user with no extended, missing, or delayed cycles."[34] Wozniak thus implies that there is no benefit to requiring, or even allowing, programmers to observe and to structure their software around refresh cycles. Making invisible a process deemed to be overly technical for applications programmers can only help them to focus more directly on their software development goals.

The sudden increase in references to transparency in 1982 appears to be an effect of the ongoing expansion of *BYTE*'s readership associated with its deprivileging of hobbyists and greater accommodation of readers from nontechnical backgrounds. Nonetheless, given that principles of transparency were often invoked to discuss system design and programming tasks, the trend in figure 1.2 requires some further consideration. When I performed another frequency count, this time checking for phrases of four words or less that incorporate variations of both "user" and "transparency," I noticed that the increase in references to transparency was significantly lower than the trend indicated in figure 1.2. To explore this difference further, I performed a similar frequency count of phrases including variations of both "user" and "friendly." The resulting count shows that references to user-friendliness were almost nonexistent prior to the 1980s but that they dramatically increased in 1982 (figure 1.3). These differences suggest that "transparent" was more often treated as a technical term to describe the design of software and hardware components and that the idea of friendliness served as a more marketable way to describe the benefits of simplified designs to the growing numbers of users from nontechnical backgrounds. While I was able to confirm these observations through close analysis of specific examples, I also noticed that there were competing definitions of "user-friendly" present in *BYTE*. The conflict between them would remain largely unresolved until the release of the Macintosh in 1984.

When we consider the market history of personal computers in the United States, both the trend visible in figure 1.3 and *BYTE*'s realignment of its editorial priorities are not surprising. Sales of personal computers increased dramatically beginning in 1980; however, as I discuss in chapter 3, this was not the result of a growing excitement over hobbyist models of computing but instead the result of the influence of rhetorics of usability that promised a new kind of "appliance computer" that would be immediately usable and useful. Some companies also advertised with business users in mind, claiming that their products would offer a cheaper alternative to those made by established office automation companies like Wang Laboratories. Sales of personal computers in the early 1980s initially increased dramatically, from approximately 800,000 in 1981 to 2.5 million in 1982, continuing to rise to 5.8 million in 1983 and 7.7 million in 1984 before plateauing.[35] Many journalists believed that the increased interest in personal computers in the United States was the result of the IBM's entrance into the market with the release of its PC 5150 in 1981. The IBM brand, they argued, was able to lend personal computing a sense of seriousness and utility that smaller, less well-known companies like Apple had not been able to.[36] IBM alone did not account for these new sales, however. Because IBM's PC used a de facto open architecture, competitors were able to release their own clones that could largely support much the same hardware peripherals and

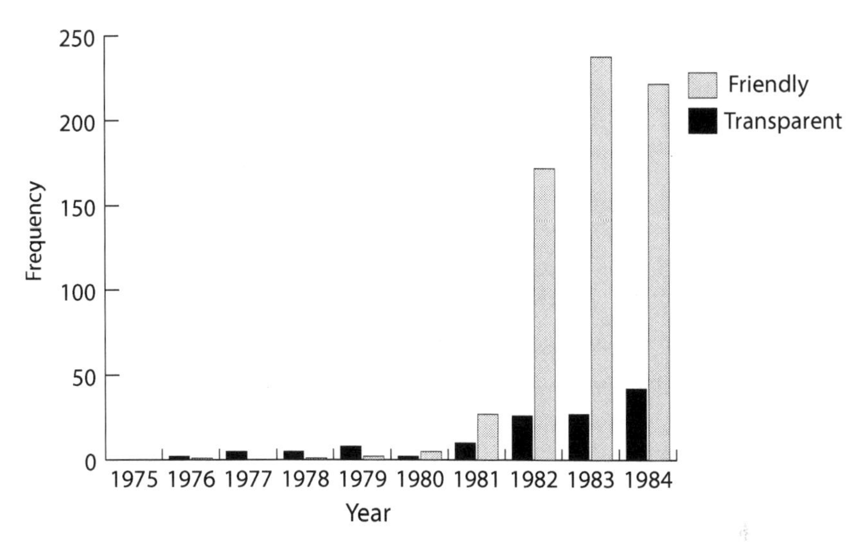

Figure 1.3. Annual frequency counts of n-grams that include "user" or "users" and either a variant of "transparent" or "friendly" appearing in *BYTE* magazine, 1975–84. An n-gram length of four was used to account for phrases like "friendly to users" and "transparent to the user." To avoid duplicate counting of matching phrases less than four words in length, the algorithm was set to ignore the next three 4-grams after a match was identified.

software applications.[37] The PC thus provided a relatively stable platform, encouraging the development of software that in turn strongly influenced public opinion about the utility of personal computing. Yet even as more and more people began to see personal computers as useful, many remained concerned that they were not usable.

At the same time that magazines like *BYTE* shifted their editorial priorities to accommodate the interests of users who were not coming to computing from technical backgrounds, they also began to talk more and more about problems of usability. Transparent design is reframed in this context as part of a technique to make computing seem less intimidating to the general public. For example, a 1981 article comparing different database management software notes that most modern applications use a storage method that is "transparent to the user," hiding the specifics of its data structures in order to provide users with a user-oriented interface that is "both more versatile and more pleasant."[38] Advertisers also sometimes included the term in their copy as part of their explanations about how their products were designed to be "friendly" to novices. Printer interface cards were marketed as capable of creating a "transparent computer-to-printer link" that allowed users to continue working while the cards' firmware invisibly handled the spooling process for them.[39] An advertisement for a programmable data analysis software package promised that "through the use of menu driven, screen oriented, tutorial response techniques . . . user acceptance will be immediate." The software "is almost totally transparent to the user, requiring no programming skill, yet achieves professional results."[40] Just as in Wozniak's discussion of transparent refresh, each of these descriptions implies that nothing of value is lost when it is hidden. References to transparency in descriptions of user-friendliness suggest that concealing overly technical aspects of computing from users would empower them, enabling them to realize more readily their computing goals. These invisible tools would not interfere or intervene in the activities of users; they would instead ensure their success.

However, as the significant differences in frequency of references to transparency and user-friendliness in figure 1-3 suggest, principles of transparency did not always inform discussions of "friendly" hardware and software. For example, some companies also described products that provided users with easier ways to find information about the system as "user friendly". This understanding of user-friendliness was often associated with the IBM PC. As I discuss in chapter 4, the PC was initially supported by a rhetoric of usability that suggested that the PC's interface would allow novice users to study and learn about computing at their own pace. User-friendly models of usability were in this context understood to be ones that would, over time, help users to develop a sophisticated and eventually expert level of computer literacy. For example, Netronics advertised its PC compatible clone as having a "user-friendly operating system that allows easy program generation and debugging. The commands include . . . a complete system test program that tests and reports the condition of ROM, RAM, cassette interface, timer, DMA controller, interrupt controller, and the communications program." The ad noted that in addition to assisting in system troubleshooting, the programs also "serve as hardware and software learning tools."[41] An advertisement for a PC-compatible modem similarly stressed that it was the most "user friendly, most reliable, and best performing modem" because it readily afforded users "total control, operation, and optioning . . . from the keyboard. A user-friendly HELP list of all interactive commands is stored in modem memory for instant screen display."[42] While this alternative definition of user-friendliness was prevalent for a time, even IBM would distance its products from it near the end of 1983 shortly after launching its "Modern Times" advertising campaign, replacing it with a rhetoric of usability that emphasized the immediate utility of its products.

The rapid increase in sales of personal computers in the United States was accompanied by a growing concern that the market's growth could not be sustained unless a significant effort was made to reduce the demands on users' time and attention. Many industry observers thus contributed to a rhetoric of usability that suggested that computing needed to somehow become less technical. Regardless of what form the next generation of interfaces would take or which company would ultimately lead the way, "the consensus," as one *80 Microcomputing* journalist reported from the floor of the 1982 Northeast Computer Show, was that "software will be virtually transparent in years to come. The microcomputer's appeal to non-technical users will increase at the expense of the hard- and software hacker."[43] Many felt that lingering concerns about the usability of personal computers were just the result of an unfocused market. In a lengthy essay published in *BYTE* that same year, for instance, Chris Rutkowski argued that computers would only become suitable for nontechnical users once designers achieved "architectural stabilization" by "design[ing] out technical choices." Computer experts, he explains, "often want to build in every conceivable option because 'you never know what the user may want to do with the system.'"[44] But instead, he argues, they should design personal computers that carried out a more limited set of specific tasks, implementing preset configurations and automating common procedures so that users could focus more narrowly and immediately on achieving their intended reason for using a computer. Through constraining design in this way, "the relationship of user and tool approaches one of *transparency*." Like many advocates of transparent design, Rutkowski points to the automobile as an example for designers to follow. Many people know how to drive a car, but few have intimate knowledge of how the drivetrain functions. This lopsidedness, he suggests, is evidence of transparency's potential for success in computing.[45] While the principles of transparent design were originally applied to narrowly defined, highly technical contexts, by 1982 many in the computing press were suggesting that they could be applied as theories of interface design as well to address growing concerns about usability among the general public. Nothing of value would be lost by narrowing the scope of potential uses for personal computers.

User-Friendliness as a Hack for Computer Literacy

Tracing the application of transparency's principles to usability reveals a pattern across the contexts I explore throughout this book. During the first decade of personal computing, technologists repeatedly reframed the sociocultural complexities of computing to define them as problems that could be solved with comparatively simple, intuitively applicable solutions. This approach to problem solving is similar to hacking in that it highlights the cleverness of a designer while also allowing them to portray themselves as committed to a socially responsible approach to computing.[46] In practice, however, designers have tended to ignore many of the sociocultural concerns that they claimed to be addressing. The promotion of transparent design at the expense of other approaches to user-friendliness serves as a compelling example of the idea that simple solutions are always the best. Despite IBM's reputation prior to the 1980s for tightly controlling its leased mainframes, its construction of user-friendliness had much in common with the countercultural visions of personal computing popularized by Ted Nelson and Steven Levy. Like Nelson and Levy, IBM encouraged an individualized relationship between user and system. For users to truly make their computers "personal," they would first need to develop their own computer literacy via a curriculum outlined in IBM's documentation. IBM's competitors, however, suggested that the kind of computer literacy celebrated in the PC's documentation was too complicated a goal to realize on a mass scale. The PC's model of usability was to them just another example of IBM's reputation for privileging computation over culture. Truly user-friendly software, they suggested, should not require users to acclimate themselves to their computers at all.

Regardless of which approach we might prefer, teaching people how to use computers was and is a complex undertaking involving many difficult choices. Scholars in rhetoric and composition have defined literacy in a variety of ways but generally agree that the scope of computer literacy extends beyond a functional understanding of personal computers. By "functional understanding" here, I mean the acquisition of skills related to the everyday use of computers to facilitate creative or productive tasks at home, school, or in the workplace. Both Cynthia Selfe and Stuart Selber have shown that computer literacy programs sponsored by the US government, businesses, and higher education have historically focused on functional understandings of computers. As Selfe argues, however, computer literacy is characterized by a "complex set of socially and culturally situated values, practices, and skills." Computer literacy is not just the procurement of technical skills; it also necessitates an understanding of the "social and cultural contexts for discourse and communication, as well as the social and linguistic products and practices of communication and the ways in which electronic communication environments have become essential parts of our cultural understanding of what it means to be literate."[47] Computer literacy is not just our ability to perform specific tasks using computers but also our ability to navigate and respond to the norms of computational environments and by extension the sociocultural contexts that computers have been integrated into. Selber suggests, too, that our ability to engage with computers critically is in many ways dependent by our functional understanding of them.[48] By limiting our understanding of computer systems to specific, prestructured tasks, the simpler models of usability we internalize to perform them and that are built on the principles of transparent design necessarily constrain our rhetorical and critical engagement with computers. Their associated rhetorics of usability invoke the idea of user-friendliness to encourage us to interpret transparency not as a constraint on our digital agency but as a relief from the unnecessary burden of engaging with computational concepts more directly. User-friendliness encourages us to accept as a condition of use that the overly—or merely—technical aspects of a computer system have no meaningful relationship to activities we seek to perform with these machines. Computation and its consequences slip away as we are pushed to focus on getting things done, making some aspect of our lives more convenient, or reveling in someone else's vision of a digital utopia.

Any potential drawbacks to a circumscribed form of computer literacy are rarely acknowledged in early computing magazines. One reason for this may be that many contributors believed in the techno-progressive narrative that has defined personal computing since the early 1970s. In her study of technochauvinism, Meredith Broussard shows how the American tech industry has been supported by an "unwavering faith that if the world used more computers, and used them properly, social problems would disappear."[49]

From a technochauvinist perspective, user-friendly approaches to design are always performed in good faith and will always, ultimately, be of benefit to the general public. Because user-friendliness is so strongly associated with a master narrative framing personal computing as an effort to democratize technology, any challenge to transparent design is received as advocating for a return to an elitist period when the use of computers was limited to those who had formal training and who worked in rigid technocratic bureaucracies. Today, the "personalness" of our computers helps to carry the idea of decentralized, democratic technology forward, directing our attention away from the fact that the software and data that flows into them come from a shrinking number of sources. Just as transparency conceals the mechanics of computation from us, so too does user-friendliness direct our attention away from the role that Big Tech's designers have in shaping the cultural, social, and political contexts of computer use.

Now that computers have become invisibly integrated into almost every aspect of our lives, our relationship to American society and its cultural institutions is determined in no small part by our ability to develop a computer literacy that conforms to the norms reified by the unseen mechanisms supporting user-friendly software. Annette Vee describes the skills and understandings we develop via personal computers as a "platform literacy"; indeed, they are now so entrenched in almost every cultural and social activity we engage in that we have reached a moment when it has become more appropriate to refer to computer literacy simply as "literacy" rather than treat it as a specialized category.[50] Computer literacy is developed through interaction with specific pieces of hardware and software and not through an engagement with some abstract, Platonic ideal of computing. According to Vee, platform literacies are also sociomaterial, changing over time in response to shifts in norms or transformations in the technological paradigms we realize our skills and contextual understanding of computers through.[51] Today, our computer literacy is being defined and redefined by small groups of designers working to further their own interests under the cover of "friendliness."

The stakes of this elite control over the hardware and software that informs our platform literacies can be illustrated by way of reference to the idea of infrastructural imperialism. According to Siva Vaidhyanathan, infrastructural imperialism is the strategy that lies beneath the tendency of Big Tech companies like Google to offer their technologies to existing social institutions cheaply or even for free. The rhetorics of usability accompanying their products promise users real and easily accessible benefits to their individual working lives and employers a measurable efficiency for their institution's operations. These free tools inform the platform literacies of an institution's members, leading those members to change how they perform their roles to better conform to their friendly models of usability. Soon, the whole institution is remade in the software's image. Ideally, the institution's administrators can be persuaded that its functioning has been improved, but even if they cannot, the cost of abandoning Google's models of usability is at this point prohibitive. It is far more expedient now to adopt Google's rhetoric to justify the changes.[52] The companies managing our everyday personal computing infrastructure like Apple, Google, and Microsoft assure us that their software is built first and foremost with our needs in mind, designed to fit seamlessly into our lives by being transparent to our intentions. We often have little choice but to accept what they say because so many aspects of our private and professional lives depend on access to their software. Yet the computer literacy afforded users on a mass scale is often distinct from that of the software's designers, who as "experts" can confidently dismiss our concerns over how their software affects us by simply stating "that's not how it works."

This inequity in levels of computer literacy is no accident. Foundational works of usability theory assert that separating the management of technology from the use of technology is a primary goal of transparent design. In *Understanding Computers and Cognition*, for example, Terry Winograd and Fernando Flores explain that a major benefit of transparent design is that it can act as an extension of managerial policy. User-friendly interfaces establish firm boundaries between domains of expertise so that the technical aspects of computer systems become "the province of the system designers and engineers" while "the user operates in a domain constituted of people and messages."[53] In practice, however, there is no clean separation between computation and culture. Safiya Noble and Tara McPherson have shown that designers inevitably embed their own assumptions about culture and society into the technologies they build. These beliefs are latent in the algorithmic systems they design, but because those systems exist in a space that their technochauvinism leads them to believe is neutral, they refuse to recognize the sociocultural nature of design.[54] Noble's work, in particular, has shown how companies like Google can readily exploit the inequity in levels of computer literacy to dismiss criticism. They respond to the idea that any cultural, social, or political influence on their work is due to us projecting our own biases onto it and insist instead that we speak to them through a more objective, technical framework. Noble has documented ways in which Google's search algorithms enact racial violence through their privileging of racist and sexist stereotypes in the results they produce. When confronted with her study, Google's engineers summarily dismissed her work, insisting that the real problem was that she does not understand how their technology functions.[55] This move is unfortunately all too common and an example of the continued prevalence of "computer crud": the insistence by all manner of technology professionals from your company's computer guy to Big Tech's top executives that computing problems can only be legitimately acknowledged when described through a supposedly value-neutral jargon. Much as Ted Nelson claimed it did in the 1970s, computer crud is today a key part of how technologists assert their authority over the sociocultural aspects of computer use.

The scholarship I have cited is part of a growing effort to hold developers of online platforms accountable, but there has been less scrutiny of designers' intentions when it comes to the personal computing software that structures our most basic interactions with digital culture. The cultural implications of operating systems and web browsers are hardly remarked on in digital media studies even though they mediate all of our computing behaviors. Consider, for example, how Google promotes individual user choice and responsibility over data privacy in the rhetoric of usability embedded in its Android operating system. Because users have the ability to set blanket device-wide or individually tailored "permissions," they need not, Google suggests, be concerned that the applications they install are using their private data in ways they might find objectionable. The permissions themselves are abstract and fuzzy, pertaining to accessing certain types of data or input sources like body sensors, calendar, camera, contacts, location, microphone, phone, SMS message logs, and storage media.[56] While permission requests from individual applications may give users a sense of why a developer would want access to their private data, the settings themselves do not reveal anything about acceptable uses for that data, which pieces of it will be used, or whether or not any data will be sent off the device. Blocking certain types of access may give users some sense of privacy; however, independent security specialists have found that even when permission is denied that many applications are still able to access most data types and input sources anyway by finding alternative storage or caching locations within the Android file system.[57] The majority of users lack the training and resources to uncover these vulnerabilities; they only know what the device's interface allows them to know. The arbitrary relationship between interface and algorithm allows designers to insist that devices behave however they see fit, in whatever manner is necessary to advance their agendas of surveillance and extraction. Users who are unwilling to risk surrendering their data have little choice but to remove applications or stop using their phone entirely, neither of which may be viable options if their working or private activities require their use.

Further, user-friendliness and transparent design have also afforded software developers the ability to manipulate models of usability on the fly. Automatic, mandatory updates provide designers with a degree of control over entire media ecosystems that could hardly have been anticipated in the 1980s, especially given that the decentralization of computing power was generally understood to be animating the growth of personal computing. Yet today's personal computers have become so radically recentralized that they are "personal" in name only. Not only do personal computer operating systems facilitate surveillance and data extraction; they also wait patiently for remote commands issued to them from their designers. When announcing Windows 10, for example, Microsoft declared that it would be the company's "last" operating system. Users would automatically receive minor updates on an as-needed basis as well as major "feature" updates twice a year that would over time make the same kinds of changes they had previously seen only after upgrading from an older operating system to a newer one.[58] All updates are downloaded and installed invisibly and automatically unless users purchase upgrade licenses that unlock additional configuration options. By default, users are only made aware of the update process when a system restart is needed and can only request to defer the restart temporarily. Users are left no option other than to trust that these continual updates will not prove disruptive. However, recent reporting has found that some updates have deleted user data or "bricked" their devices, leaving Windows 10 unusable until the installation can be repaired.[59] Many users have also expressed concern that even when these automated updates work correctly they are making undesired changes to their devices.[60] Because updates are mandatory, automated, and increasingly performed without asking for our consent, there is no way to contest Microsoft's continued intervention in our use practices apart from rejecting Windows itself. For many users, such a rejection would be akin to abandoning personal computing entirely. The choice is not between black-box and open technology. Instead, it is between either accepting Microsoft's "user-friendly" update policies or giving up the software we have come to depend on for work and leisure that must be run on Windows.

Conclusion

As I have shown in this chapter, the transfer of a technological problem-solving strategy onto complex questions about computer literacy in part accounts for our now conventional understanding of user-friendliness. The solutions software designers offer us sustain an inequity in computer literacy, constraining users by allowing them to engage with technological concepts only in comparatively limited ways. This inequity further allows technologists to position those aspects of computing that they deem to be merely technical in nature as existing outside of culture. Any concern expressed by users over the sociocultural implications of the algorithms that designers represent as operating in a neutral, objective manner can easily be dismissed as misguided, as most users lack the computer literacy necessary to criticize the computing industry in ways that technologists will accept as legitimate. For this reason, critics and historians must work to develop a counter rhetoric of usability that opens up a space for popular critique of the self-evident nature of transparent design that cannot be dismissed as a call for a return to elite computing. Regardless, we cannot "return" to elite computing because we never left it. The contemporary push toward cloud computing, for example, is not new. It is the culmination of calls made beginning in the mid-1970s for the elite management of core aspects of personal computing technologies in the interest of expediency. We began on the path back toward centralization almost immediately after personal models of usability were first realized.

To push back on the idea that transparency comes for free, we must reject the problematic aphorism implied by it. What we don't know *can* hurt us. Transparency comes for free only if we accept the technocratic assertion that those hidden elements of a computer system are merely technical in nature and thus culturally inconsequential. We should also regard the explanations of function provided by an interface with skepticism, especially when the rhetorics of usability associated with computing technologies outwardly promise to provide some radically simple solution to a problem in our private or working lives. Even if an application or device appears to function as promised, the very fact of a gap between interface and algorithm can mean that it is also working at cross-purposes to accomplish an unseen goal of its designer. Many scandals involving Big Tech in recent years have in part been the result of designers exploiting users' inability to see what is happening beneath the interface. What is one to do when something as seemingly benign and inconsequential as a weather app is revealed to actually function as a data collection mechanism for a machine- learning-powered system for targeted advertising?[61] Or when a new operating system update promising increased security and stability secretly slows down your phone?[62] The computer literacies we develop through our continual use and exposure to user-friendly, transparent software does not permit us to verify that an application is not doing more than it promises. We are told that to be responsible digital citizens that we should make informed choices about the media we consume and the technological systems we engage with. Yet if interfaces hide important details about their function from us, it becomes very difficult if not impossible in practice for most of us to make informed choices about how we want to integrate technology into our lives. Instead, we come to rely on companies like Apple, Google, and Microsoft to make those choices for us.

Today's user-friendly, ultra-transparent computers are anything but personal. The ability of hardware and software developers to push automatically installed updates allows them to regularly intervene in and modify our computational practices and thereby continually normalize our use behavior. Rhetorics of usability press users to internalize and view as desirable the models of usability that designers embed in their software. Yet developers now also have the power to alter their models of usability at will and so can subvert any attempt to challenge, repurpose, or otherwise pursue alternative models with a given technology. As users move to socially construct the technology before them in order to engage in use behaviors that designers did not anticipate or otherwise disapprove of, software developers can simply push out an update that blocks that behavior. Digital media critics have described "glitches" as a path to resistance to technological power in the sense that they can be exploited by users to subvert the limits imposed by models of usability or that users encountering them in digital art will begin to realize the arbitrary limits on their digital agency imposed by transparent design.[63] As I discuss in greater detail in chapter 5, foundational literature in the field of human-computer interaction provides a scientific and philosophical justification for calls to construct technologies that passively discourage such reflection. The earliest, formal articulations of transparent design as a distinct theory of usability, in other words, position the reflective awareness that follows a moment of glitch or breakdown as something to be avoided at all costs. In seeking to compel users to internalize idealized use behaviors, software developers thus constantly search for and eliminate glitches or gaps in function that could perhaps be exploited through the constant procession of updates. This problem is not theoretical or speculative: it is already happening. Manufacturers of game consoles, for example, regularly crack down on hardware and software modifications that permit anyone other than an authorized development partner to design software for their platforms.[64] They also continue releasing updates on platforms that are no longer officially supported for the sole purpose of fixing exploitable glitches.[65]

If rhetorics and models of usability are on their surface about explaining computing, providing mental models, and encouraging specific use behavior, we must recognize that they also fulfill a political function. As I show in the chapters that follow, our popular understanding of the history of personal computing in the United States has been strongly influenced by rhetorics of usability that were devised to support the business models of hardware and software developers. Ensuring that hardware and software are received as natural or intuitive minimizes the continued work that developers must do to build and maintain their preferred interpretations of their technology. In the next chapter, I examine how the idea of a personal computer revolution as a countercultural movement that sought to create inclusive models of usability was a narrative adopted during moments before the formation of technological standards and then quickly abandoned as hobbyists found that their political commitments proved technologically inexpedient. While popular histories of the period often frame 1970s counterculture as an animating force driving early personal computer development in the United States, returning to early newsletters and magazines shows that only a minority of influential hobbyists embraced the countercultural vision, and many of those were quick to abandon that vision by the end of the 1970s in favor of a new rhetoric of usability that suggested that computers should be more like appliances than tools for exploration. Picking up on several talking points I have introduced in this chapter, in the next three I examine the rhetorics of usability associated with the first series of commercially manufactured personal computers in the United States from 1977 to 1984. In the course of trying to resolve conflicts among varying rhetorics of usability, many industry observers came to accept that transparent design was the best and eventually only way to realize user-friendliness's promise of making computers immediately usable and useful.

2

The Sources of the Personal Computer Revolution

The influence of user-friendliness on American views of digital culture cannot be understood apart from narratives of a personal computer revolution. While the technical term for the desktop-sized machines that first became available in 1975 was "microcomputer," by 1976 the term "personal computer" had become more common. The term suggested that computers could be for "personal-use" and could "increase personal freedom."[1] Yet, in many ways these two ideas are in tension with one another, as the former has come to be associated with models of universal usability and the latter with models of decentralized usability. Today's rhetorics of user-friendliness are tangled in both, leading to conflicts and contradictions between the language and materiality of computing.

In this chapter, I examine how the concept of personalness has served and continues to serve as an ethic of expediency that hardware and software designers have appealed to since the 1970s to justify their technical practices. As defined by Steven B. Katz, an ethic of expediency is a form of deliberative rhetoric within which "the only ethical criterion necessary is the perceptible movement towards the technological goal to be achieved."[2] Ethics of expediency conflate technological progress with social and political progress and serve to conceal or to legitimate any harms done in the service of a technological goal. Advocates of a personal computing revolution insisted that decentralizing computation and putting computers into the hands of as many people as possible would make the United States more democratic; however, they quickly encountered several social and technical obstacles to realizing that vision. When facing these problems, they frequently opted for technologically expedient solutions. Because these solutions conflicted with their stated political commitments, they revised their rhetorics of usability at key moments to convey the sense that these solutions were indeed necessary to realize the larger goals of the revolution. One major outcome of these conflicts is an early acceptance of centralization via the private management of key components, after which, influential hobbyists began to position themselves as a new elite, arguing for universal usability and claiming that it could only be realized through their benevolent management of computation. Here I trace the revisions and adjustments that these writers made to their rhetorics of usability in the mid to late 1970s, highlighting how they worked to reframe the revolution in ways that later writers would draw on to support their arguments in support of transparent design.

Many details of the personal computer revolution are familiar to us today, playing an important role in the innovation mythology from which Silicon Valley derives its cultural authority. Small groups of computing enthusiasts, who referred to themselves as "hobbyists" and who shared a belief in personal freedom and self-empowerment through technology, came together in the San Francisco Bay area to pool their skills and explore alternative models of usability made possible by microcomputers. Many of the first and still most influential iterations of this narrative of revolution were written by technology journalists such as Paul Freiberger, Michael Swaine, Steven Levy, and John Markoff.[3] Yet there always have been outside observers skeptical of this framing. Writing in 1988, Bryan Pfaffenberger argued that the countercultural views of these hobbyist revolutionaries were distinct from those found in other, contemporaneous political movements. Many influential hobbyists, he explains, largely accepted "the dominant value system" of professional engineering culture. Pfaffenberger implies, in other words, that they were privileged white men who had avoided participating in Vietnam and were largely unaffected by and uninvolved with the struggles of antiwar groups, the civil rights movement, feminist politics, the LGBT community, and other oppressed peoples within the United States. Pfaffenberger goes on to suggest that they adopted a countercultural self-representation primarily because they were "marginal to the corporate system in some way, often because (like [Steve] Wozniack or Captain Crunch) they hadn't completed their engineering degrees, or (like Lee Felsenstein) they could not stomach corporate authority—or, like Bill Gates in the mid-1970s, they were far too young to enter it."[4] More recently, Elizabeth R. Petrick's review of early hobbyist newsletters has shown that their countercultural rhetoric was often mixed with and superseded by discussions of entrepreneurial enterprise.[5] Given the prominence that authors like Freiberger, Swaine, Levy, and Markoff ascribe to the role of countercultural politics in early personal computing, studies like Pfaffenberger's and Petrick's serve as calls to return to and reevaluate the sources that informed the narrative framings that Silicon Valley continues to draw on to explain the cultural significance of its work.

Recent reconsiderations of personal computing's politics have identified a disconnect between hobbyists' stated values and their actual practices. The most significant challenge to conventional accounts of the politics of personal computing can be found in Joy Lisi Rankin's study of networked and timeshared computing in the 1960s and 1970s. In her conclusion, Rankin argues that the events of the personal computer revolution brought about an end to a "golden age" during which students, teachers, and enthusiasts working in universities, schools, and other educational institutions defined new, inclusive "computing communities" by writing their own novel software and exploring the potential for art and activism via computers.[6] The practices that Rankin documents throughout her book very much resemble the practices that early hobbyists claimed to value. Yet Rankin concludes that the introduction of individually owned personal computers led to the foreclosure of possibilities for social change through computing, signaling the end of a model of usability based on the free, unstructured exploration of computation.[7] Similarly, Luke Stark's examination of the flyers, advertisements, and other promotional material that hobbyists developed to promote their start-ups to one another suggests that by the late 1970s they were more interested in having their own technological prowess acknowledged than in social change.[8]

Returning to texts read or written by early personal computing enthusiasts, I trace how the idea of personalness as an ethic of expediency induced hobbyists to continually revise their rhetorics of usability in such a way as to suggest that the personal freedoms they pursued could not be realized through the radical decentralization of computing hardware and computer literacy but only through the privatized control of key technologies. These revisions were part of a process of recentralization understood as necessary to realize a universal model of usability. To trace these revisions in hobbyists' rhetorics of usability, I take as my starting point Levy's *Hackers: Heroes of the Computer Revolution*, one of the earliest books written about the hobbyists' revolution. Many ideas and rhetorical frameworks from Levy's book are still prevalent in popular discourse today, evoking the master narrative that seems to speak from the presumed center of Silicon Valley culture that Susan Leigh Star describes.[9] In his book, Levy posits the existence of a "hacker ethic" and insists that hobbyists generally shared uniform political commitments as a result of their exposure to the writings of Stewart Brand, Ted Nelson, and Ivan Illich. Despite an apparent unity of purpose, however, I show that there were significant disagreements across these three writers that Levy fails to address, specifically with regards to the power dynamics between users and designers. These disagreements help to illustrate how the idea of personalness became a key principle in later rhetorics of user-friendliness, especially in the way that it helps to conceal the consequences of designers' decisions from users when those users are interacting with transparent interfaces. Whereas Brand and Nelson suggest that users could pursue strategies of decentralization in order to establish individualized relationships with technology that were free of the influence of its designers, Illich insists that institutional power is inevitably embedded in technology as a product of its design. The continued durability of the master narrative of the hobbyists' revolution is due in no small part to the way that today's rhetorics of usability leverage the idea of personalness to discourage support for the kind of public engagement that Illich described as necessary to produce and maintain convivial technologies.

After discussing Levy's problematic synthesis of Brand, Nelson, and Illich, I then turn my attention to writings authored by hobbyists in order to reconsider the extent to which Brand's, Nelson's and Illich's ideas are present in early users' group newsletters. Reviewing material from the *People's Computer Company* and the *Homebrew Computer Club Newsletter*, I argue that while some members appear to embrace many of the technological practices Brand, Nelson, and Illich elucidate, few demonstrate any direct engagement with or otherwise explicitly acknowledge the broader political theories or concepts they describe. This lack of political engagement is especially evident in publications not associated with the narrative that Levy and others popularized, like the *Mark-8 Users Group Newsletter* and *Computer Notes*. In light of this, I argue that for many of the hobbyists participating in the early personal computing revolution, the only goals that mattered were technological. The antiestablishment politics present in some hobbyist publications were a first attempt at articulating responses to the social and material problems that blocked them from their technological goals. Yet as the social and material contexts of computing changed, and new possibilities for practice emerged, hobbyists were open to alternative configurations of technological power. As I show, cultural moments like the Bill Gates's "Open Letter to Hobbyists," the first West Coast Computer Faire, and Apple's early advertising campaigns illustrate how the new elites who emerged promising to support models of universal usability were received favorably because they offered expedient solutions to the social and material problems that radically decentralized models of usability struggled with.

Hackers, Hobbyists, and the "Heroes" of the Personal Computer Revolution

Throughout his book, Levy argues that everyone who contributed to the personal computer revolution shared a set of core beliefs. This "hacker ethic" is defined by the following principles: a need for unlimited and total access to computers, a desire for complete freedom of information, an inherent distrust of authority, a support for meritocratic social recognition, a conviction that all technology can be used creatively, and a belief that computers can directly improve people's lives.[10] Key to each of these principles is the idea of a decentralized model of usability. Even though Levy portrays these hackers as each pursuing their own, personal relationships to computing, he also argues that they shared a commitment to the hacker ethic because it was "the computer that did the converting."[11] During the late 1950s and early 1960s, the first hackers realized that computers had no politics in and of themselves. While they could be used to oppress, they could also be used to promote free expression and creative thinking. The games and experiments these hackers engaged in after hours in academic computer laboratories were a challenge to oppressive models of computing and thus would necessarily one day improve the world. If outsiders did not share this view of technology, it was because they tended to view technology through the rigid rules and prescribed practices of technocratic bureaucracies like IBM. Levy's hackers viewed these bureaucracies as always self-serving and suggested that all the rules they imposed on access were part of a broad strategy of centralized management that allowed them to maintain a monopoly over computing power. By contrast, wherever free, unstructured access to computing was available, people would soon develop a commitment to "sharing, openness, decentralization, and getting your hands on machines at any cost to improve the machine and improve the world."[12] According to Levy, the electronics hobbyists of the 1970s were quintessential hackers. The computer kits they experimented with in their garages and basements were part of an ongoing movement to promote decentralized models of usability and similarly encouraged the free and open sharing of information about the technologies they developed via clubs and newsletters.

Levy's hacker ethic is thus a counter rhetoric of usability that reframes the status quo models of usability that were implemented in university and corporate laboratories before the mid-1970s as antisocial. Further, Levy's hackers feared that in the absence of an alternative model, all access to computing would be dictated by "arbitrary rules" devised solely to "consolidate power" over users.[13] In addition to being known for its dominance of mainframe computing, IBM was also infamous for its conformist culture and for the leasing agreements it used to keep knowledge about its products secret from competitors and clients alike. One of the focuses of the United States' antitrust suit against IBM was the company's practice of "account control." For decades, IBM had permitted the use of its machines only through leases rather than through sales. Initially, account control served as a kind of "security blanket" for new customers as they were "introduce[d] to the mysterious new machines." IBM executives and engineers would handle the planning, construction, and maintenance of the hardware, along with software development, design of a data processing strategy, and training in operation of the software on behalf of the client. Clients never saw a bill for account control as it was factored into the cost of their hardware lease. In practice, however, account control meant that "what the customer knew about data processing was the company line," as IBM's engineers essentially made all decisions regarding how their computer systems would be used.[14] Corporate clients were encouraged to see computing as a service that was contracted out, ideally a seamless extension of existing management policies and not something that executives were required to concern themselves with beyond incorporating the presumably objective and trustworthy data generated by the machines into their decision making.[15] As part of their counter rhetoric of usability, Levy's hackers insisted that, in an increasingly computerized society, this kind of information asymmetry was a threat to individual freedom. Thus, any alternative model of computing would have to be driven by a concerted effort to decentralize access so that complete and accurate information about the design and implementation of computer systems would be freely available.

Levy's hackers were concerned that IBM's dominance of corporate computing would soon extend to all shared computing access sites. In American culture, they felt, "the words 'IBM' and 'computer' were synonymous."[16] They viewed the comparatively lax rules governing access to the university laboratories they worked in as a slow creep toward the kind of total control that IBM commanded. Although there were several large computer manufacturers operating in the United States by the early 1950s, they were quickly disadvantaged in the market by IBM's defense contracts, which allowed the company to grow at a much faster pace. Since the 1940s, IBM had been regarded as "a stodgy company shackled to obsolete technology," but by the 1960s, IBM controlled 85 percent of the United States computing market.[17] Even though Sperry could boast that its UNIVAC line of mainframes had a number of technological advantages that allowed for more flexible applications, IBM's salesmen stressed immediate—and quantifiable—benefits to their customers: "Their computer would get the payroll out two days early and save vast sums of money in the process; and there was no question whose argument was more persuasive."[18] Levy's hackers feared, in other words, that in the absence of any viable alternative to IBM's rhetoric and model of usability that anyone who worked in computing would end up likely working for IBM, for a company that did business with IBM, competitors who tried to style themselves after IBM, or in military and university laboratories that had their own equally arbitrary sets of rules.

The way these early hackers leveraged their ethical principles in a world that was against them, as Levy describes it, exemplifies several aspects of Katz's ethic of expediency. Most significantly, Levy positions the pursuit of free access to computing power as a "necessary good that subsumes all other goods and becomes the basis of virtue itself."[19] Beyond opposing heavily bureaucratized institutions, Levy's hackers were not responsive to the broader political implications of computers. In one of the more revealing moments in his book, Levy describes a period in 1969 when protestors were holding antiwar demonstrations on MIT's campus. Here, the hackers were forced to recognize that "all the lab's activities, even the most zany or anarchistic manifestations of the Hacker Ethic . . . [were] paid for by the same Department of Defense that was killing Vietnamese and drafting American boys to die overseas." Rather than abandon their laboratory, they responded by claiming that their work was misunderstood. Dismissing the possibility that anything they created though their after-hours experimenting could be co-opted by the US military, they concluded that their hacking had shown the positive potential of computing. The furor outside their lab did not owe to the computers themselves but to the fact that the public perception of computers had been shaped almost entirely by "brain-damaged, bureaucratic, batch-processed mentality of IBM." While before the demonstrations, many of Levy's hackers balked at the laboratory's locks and passwords, most afterward accepted them as necessary to preserve their work "and chose not to view the locks as symbols of how far removed they were from the mainstream."[20] This incident is representative of repeated compromises made in the interest of technological expediency. Refusing to modify their technological practices, Levy's hackers had to reframe the politics of their laboratory by asserting both the ideological neutrality of technology and the virtuousness of their work. Moments like this one show how the hackers and hobbyists of the first decade of personal computing were always in practice concerned first and foremost with technological expediency regardless of how they explained the cultural, social, or political significance of their work.

Stewart Brand and Technological Empowerment

Brand's influence on Silicon Valley culture is well documented. Yet, as Fred Turner observes, Brand was largely absent from American counterculture during most of the 1970s and early 1980s after his *Whole Earth Catalog* ceased publication. Levy, however, was a fan of Brand's work and connected with him in 1984 after completing work on his book to help organize a conference for hackers, many of whom he had written about and would go on to become Silicon Valley luminaries.[21] Prior to the 1980s, Brand's only major publication on personal computing was an article titled "Spacewar: Fantastic Life and Symbolic Death among the Computer Bums," published in 1972 in *Rolling Stone*. Linking ideas from his activism to the hackers he observed at Stanford and at the Xerox Palo Alto Research Center, Brand produced one of the first published versions of a rhetoric of usability that associated decentralized access to computing with personal freedom. The bulk of Brand's article is a profile of the late-night hackers. Despite their differences, both groups had come to believe that computers could be designed to serve "primarily as a communication device between humans" and put to use in pursuit of "human interest, not machine." Brand also comments that this type of human-centered computing "function[s] best on stand-alone equipment."[22] In drawing a connection between his own countercultural politics and the late-night hackers, Brand lays the groundwork for Le y's hacker ethic and suggests that computers could be reinvented into tools for positive change if they were dissociated from the narrow-minded institutions they were now commonly found within. The kind of free access that the hackers enjoyed after hours would require that their management be decentralized. Once people were free to explore computation on their own terms, computers could be put to use in pursuit of socially beneficial purposes rather than just those that academic, corporate, or military projects deemed worthwhile.

Many of the ideas that Brand here associates with decentralized computing are extensions of the practices he had previously promoted through his *Whole Earth Catalog* beginning in 1968. Turner notes the catalog's cut-and-paste style was a product of Brand's politics, not wanting himself to become the kind of central authority that he opposed. In his role as editor, Brand exercised a nonhierarchical editorial policy, incorporating any contributions he received from readers that he or his coeditors deemed useful.[23] The catalog's pages are thus a bricolage of book and magazine covers, handwritten notes, copies of typed pages, photographs, and advertisements. The catalog's introduction, however, is safely attributable to Brand himself:

> We *are* as gods and might as well get used to it. So far, remotely done power and glory—as via government, big business, formal education, church—has succeeded to the point where gross obscure actual gains. In response to this dilemma and to these gains a realm of intimate, personal power is developing—power of the individual to conduct his own education, find his own inspiration, shape his own environment, and share his adventure with whoever is interested. Tools that aid this process are sought and promoted by the WHOLE EARTH CATALOG.[24]

Large social institutions, he argues, put the preservation and expansion of their own power above the interests of the people they allegedly serve. From this perspective, the seemingly impossible problems facing American society during the 1960s were the result of the ways that education, training, and other potential forms of intellectual development were all structured to sustain the institutions that provided them rather than empower the individuals who moved through them. Brand thus calls for a decentralization of knowledge, suggesting that the catalog is a tool that would allow its users to remove technologies from the influence of the hopelessly corrupt institutions that produced them. Once separated from their original contexts of use, these tools could then be repurposed in pursuit of enlightenment and self-empowerment.

Through the catalog, Brand promotes an ethic of expediency that like Levy's hacker ethic assumes that the pursuit of personal technological and spiritual power would result in broad social change. The *Whole Earth Catalog*'s explicitly stated goal is to provide readers with access to tools, broadly defined, that they could use to empower themselves to explore alternative social structures intended to correct the problems he and others saw across the United States. The "tools" for self-enrichment that the catalog provides access to include ideas, books, techniques, camping equipment, farming implements, and decorative objects. The first issue of the *Whole Earth Catalog*, for example, prints excerpts from essays and poems alongside addresses and phone numbers that readers could use to purchase textbooks, vinyl records, kerosene lamps, hiking boots, hypnotism manuals, and pillows for meditation. Each item is accompanied by a review or submission statement, explaining how it had or could be used along with a price and contact information for a manufacturer or distributor from whom the item could be purchased. Turner describes Brand and his reader-contributors as participating in a "New Communalist" political movement. These New Communalists saw tools not as objects but as material embodiments of self-enrichment processes; for them, learning to use a tool was not so much about accomplishing a task or developing a new skill as about being able to see the world from an alternative perspective.[25] As Turner notes, Brand's catalog drew on the writings of Buckminster Fuller to promote a new mythos for American consumerism, one in which the products of modern industry could be wholly dissociated from the ideological commitments, cultural biases, and political strategies of the scientific and industrial institutions that produced them. Talented individuals could exit society, taking with them the tools they needed from the catalog and repurpose those tools to found small communities dedicated to improving the "whole person." Although each participant would ultimately use the catalog to define their own path to self-empowerment, the members of these communities would be engaging collectively in a project to improve the "whole population."[26]

Much like Levy's hackers, the New Communalists following Brand's principles made little effort to engage with the broader cultural, social, and political issues that other countercultural groups in the United States were addressing. As Turner explains, the communes that formed through Brand's influence were largely unwilling to engage with inequities related to race, gender, or sexuality among their members. Like Levy's hackers, the *Whole Earth Catalog* also ignored the Vietnam War except "insofar as it generated new 'tools' for personal transformation at home."[27] The New Communalist movement eventually collapsed because it lacked "structures of governance and structured ways of making a living—the very institutional elements of social life that many New Communalists had hoped to avoid."[28] In short, the activists who followed Brand's prescriptions found that the social change they imagined could not be enacted, even on a small scale, without some kind of formal governance to protect the values they held dear.

Unable to reconcile their rejection of social institutions with this need, most communes, Turner notes, were short lived. In later years, the rhetoric of Brand's New Communalists was ultimately taken up by right-wing politicians and libertarian technocrats who reframed the personal computer revolution as proof of the threat that regulation posed to America's information infrastructure.[29] A representative example of this reembrace of Brand's ideas is a 1994 statement by Esther Dyson, George Gilder, George Keyworth, and Alvin Toffler titled "Cyberspace and the American Dream: A Magna Carta for the Knowledge Age." The document describes Brand's ideas as the driver of a "renaissance of American business and technological leadership," arguing that "in the transition from mainframes to PCs, a vast new market was created" that was "characterized by dynamic competition consisting of easy access and low barriers to entry" making it possible for "start-ups by the dozens" to take on "the larger established companies" and win.[30] Silicon Valley succeeded, they maintain, because its innovators were able to survive outside of the reach of outmoded corporate bureaucracies and government regulations alike. Brand himself did not seem opposed to this interpretation. A year after its publication, he joined its authors along with Newt Gingrich, John Perry Barlow, and representatives from major tech companies like Microsoft at a conference to continue discussing the essay's ideas.[31] Although Brand skipped out on the first decade of personal computing, the rhetoric of usability he developed in his catalog and tried briefly to apply to computers for *Rolling Stone* would eventually find new life as part of the ethic of expediency driving the innovation mythology of Silicon Valley's start-up culture.

Ted Nelson's Push to Decentralize Computer Literacy

While Brand effectively sat out of the first decade of personal computing, many of his ideas were applied to hobby computing by Ted Nelson. Levy describes Nelson's *Computer Lib/Dream Machines* as "a virtual handbook to the Hacker Ethic."[32] First printed in 1974, the book's page layouts resemble the cut-and-paste style of the *Whole Earth Catalog.* Nelson divided the book into two halves. The first half is a mixture of his thoughts on the roles computers were playing in American culture, summaries of basic computing concepts, and—as with Brand's catalog—lists of addresses, phone numbers, and other information that readers could use to get educational materials to teach themselves about computers or secure accounts with local, time-shared computer centers. The second half is an in-depth discussion of the ways that he believed hypertext could support self-directed education and new forms of literary expression. While the majority of *Computer Lib* is devoted to helping readers develop a functional understanding of core computational concepts, Nelson also at times treats computer literacy similarly to what Stuart Selber describes as a "digital multiliteracy" by recognizing that functional understandings of computers necessarily inform our ability to think critically about and express ourselves with them.[33] Nelson argues, for example, that "any nitwit can understand computers, and many do." However, he continues, the people that design and manage computers are reluctant to share information about them with the public. They exercise a monopoly over computer literacy to promote the false impression that computers are only usable for unimaginative purposes.[34] Nelson thus asserts that decentralizing computer literacy is about more than merely liberating computers from the control of technocrats; decentralizing computer literacy is also, just as Levy suggests in connection with the hackers he discusses, crucial for sustaining democracy in an increasingly computerized society. Computer literacy will expand our ability to think creatively, navigate culture, and upend the social problems caused by the inequitable distribution of computing power.

Nelson spends a significant amount of time in *Computer Lib* describing and critiquing the "computer priesthood." Just as Levy's hackers and Brand's

New Communalists did, Nelson views computers as ideologically neutral. Nelson offers a number of possible motivations behind the priesthood's purposeful occulting of computation but ultimately argues that the gap in computer literacy between members of this priesthood and the general public is part of a strategy to seize power within those social institutions that have come to rely on computer systems to manage themselves. Elite technocrats purposefully cultivate a popular image of computing as "cold immaculate, sterile, 'scientific,' oppressive. Some people flee this image. Others, drawn toward it, have joined the cold-sterile-oppressive cult, and propagate it like a faith." If computers had a reputation for being tools of oppression, Nelson argues, it is not the fault of the machines but of the people who use them to manipulate others. Members of the priesthood can do as they please, quickly defusing objections from laypersons with the magic words that "it is the computer's fault." It is time for readers, he continues, to "stop being mad at 'computers' in the abstract, and start being mad at the people who make inconvenient systems. It is not 'the computer,' which has no intrinsic value or character, which is at fault."[35] Developing computer literacy is thus not just about using computers. Nelson insists that "you can and must understand computers now" because mass computer literacy is fast becoming a political imperative.[36]

In addition to arguing for the decentralization of computer literacy, Nelson also stresses the importance of making computers accessible outside of institutionally controlled contexts. In an increasingly computerized society, decentralizing computer literacy and ensuring free access to computing power are portrayed by Nelson as not just as worthy political goals but the only goals worth pursuing. This positioning is evident in the sense of urgency Nelson conveys, like Levy's hackers, in regard to IBM:

> Through various mechanisms, [IBM] seems to enforce the principle that "Once an IBM customer, always an IBM customer." With an extraordinary degree of control, surely possessed in no other field by any other organization in the free world, it dictates what its customers may buy, and what they may do with what they get. More than this: the exactions of loyalty levied upon IBM's customers are similar, in kind and degree, to what it demands of its own employees. IBM makes the customer's employees more and more like its own employees, committing them as individuals, and effectively committing the customer that buys from them, to IBM service in perpetuity.[37]

Outside of the company, there were increasingly "whole colonies of users" who "mold[ed] themselves in its image, so that around IBM computers there are many 'little IBMs,' full of people who imitate the personalities and style of IBM people."[38] To prevent users from being absorbed into "the IBM way," computer literacy and access to computing technologies would have to be radically decentralized. Effecting such a decentralization would allow the public to cut through "computer crud": the technical jargon and dizzying rhetoric that Nelson claims the priesthood purposefully leverages to keep computer literacy inaccessible to outsiders. "So serious and abysmal" is the public's "confusion and ignorance," Nelson argues, that "anything with buttons or lights can be palmed off on the layman as a computer. There are so many different things, and their differences are so important: yet to the lay public they are lumped together as 'computer stuff,' indistinct and beyond understanding or criticism."[39] But beyond explaining how decentralized models of literacy and access would help in resisting IBM's hegemony, Nelson does not offer any specific explanation as to how his technological philosophy would address other cultural, social, or political concerns. Like Brand, he assumes that individual empowerment, if realized on a wide-enough scale, would necessarily translate into social progress.

When it comes to realizing the kind of widespread computer literacy necessary to secure democratic freedoms in a computerized society, Nelson cautions his readers to develop their skills away from formal institutions. To avoid exposure to the oppressive influence of IBM or its imitators, Nelson instructs readers to engage in self-study and to cultivate individual perspectives on the uses for and consequences of computers. Nelson is also deeply cynical about higher education curricula and academic computing groups. In most places, he remarks, the "actual alternatives" to self-study are "fairly dismal." At universities, computer courses are taught "with a mathematical emphasis at the start" for the purposes of "cut[ting] down enrollment, since they're not setup to *want* to learn about computers." Universities are too full of people with an IBM mindset, concerned with maintaining the "status" of computers as something only for "students with 'logical minds.'" Community colleges and trade schools, he continues, are no better. They "tend to prepare students only for the most humdrum business applications" with an emphasis on "programming in the COBOL language on IBM business systems." For Nelson, high school clubs and children's programs seem to have the most promise, as they often are not intended explicitly for professional development or preprofessional training.[40] Stressing the importance of decentralization and individualized access, Nelson suggests that any institutionalization of computing is a threat to democracy. People must be free to develop skills, knowledges, and practices on their own if we are to have a future free of the influence of IBM and other oppressive technocracies.

Reading Brand's and Nelson's work in relation to Levy's highlights the ways that his hacker ethic privileges the freedom side of personalness. The hacker ethic represents decentralized computer literacy and unstructured access to computing as the basis of virtue itself. Once both are realized widely enough in American culture, all misconceptions about computing will be erased. The only way to check the power of the institutions that leverage those misconceptions for their own benefit—ostensibly the source of most, if not all, the social and political problems of the late twentieth century—is to break their monopolies over hardware and knowledge. Yet neither Brand nor Nelson evince any interest in community engagement, in working for change within existing institutions, or in pushing for any sort of public regulation of technology because all of those, in their view, would limit our freedom to discover our own unique uses for technology. The shortcomings of Levy's hacker ethic are also evident in Brand's and Nelson's assumption that technology can be wholly divorced from the politics of their creators. Nelson never stops to wonder, for example, whether the computers we use are structured so that the use behaviors necessary to operate them tacitly support at least some of the institutional norms he views as threatening. Even as Brand and Nelson argue that free access to technological tools would allow us to adopt new perspectives and expand our understanding of the world, they do not consider whether the new perspectives afforded by technology might be ones intended to cultivate a worldview that benefits their designers.

Ivan Illich and Convivial Tools

While Nelson and Brand are often cited in histories and critical examinations of digital culture, Illich's influence is less well acknowledged. However, at least two members of the West Coast hobbyist community who feature prominently in Levy's book, Fred Moore and Lee Felsenstein, were familiar with his work and tried to bring his ideas into their writings and public remarks about the politics of computing.[41] Additionally, passing references to Illich's description of conviviality are not uncommon in early literature on usability. Donald Norman, for example, would later describe the goal of his new approach to "cognitive engineering" as developing "convivial tools" that would allow people to experience a sense of pleasure during use.[42] Like Brand and Nelson, Illich is critical of the institutional exercise of technological power and is excited by the potential for users to find creative, new uses for modern technology; however, the rhetoric of usability he develops in response to this concern is quite different from theirs. Whereas both Brand and Nelson advocate for decentralization by calling on readers to seek an outsider position, Illich argues that the oppressive potential of modern technology can only be checked through a collective commitment to the public accountability of the elites who control its design and implementation. This aspect of Illich's thought, especially the way he repeatedly cautions his readers to avoid prioritizing technological expediency, is not well represented in those moments when hobbyists do engage with his ideas.

Throughout his career, Illich was broadly concerned with how power spreads outward into Western culture from professional and civic institutions. However, the hobbyists who took up his work were concerned specifically with those writings discussing the relationship between modern industry and individual autonomy. Published in 1973, Illich's *Tools for Conviviality* appears in certain respects to have much in common with arguments made or implied by Brand and Nelson. For example, Illich defines "conviviality" as the "individual freedom realized in personal interdependence and, as such, an intrinsic ethical value." A convivial society is one that promotes the "autonomous and creative intercourse among persons, and the intercourse of persons with their environment; and this in contrast with the conditioned response of persons to the demands made upon them by others, and by a man-made environment." Industrialization, in other words, has empowered us by providing technologies that free us from many daily demands on our time, increasing our individual autonomy. Yet there are also significant points on which they diverge. Illich explains that while "modern science and technology can be used to endow human activity with unprecedented effectiveness," promoting creative intercourse in ways not previously possible, they are not in themselves sources of progress: "In any society, if conviviality is reduced below a certain level, no amount of industrial productivity can effectively satisfy the needs it creates among society's members." Conviviality is not something that a given technology is inherently capable of supporting; rather, conviviality can only be achieved through supportive social structures that represent a collective commitment to supporting human autonomy and creativity.[43]

Illich further argues that many industrial technologies actually hinder conviviality because they cannot be separated from the self-serving intentions of their designers. Most industrial technologies, he explains, are accessed through an infrastructure that normalizes our use behaviors or are designed in ways that enforce specific limits on user agency. Often, industrial technologies constrain the ability of the average person "to enrich the environment with the fruits of his or her vision" because "they allow their designers to determine the meaning and expectations of others."[44] Automobiles, for example, are not convivial tools because they have "created more distances than they helped to bridge."[45] Cars are presented to us by their manufacturers as a way to increase personal mobility, giving people the power to relocate themselves for any purpose, quickly and cheaply. Yet automobiles have by and large been designed primarily around a "capsule" model of transportation that favors the small vehicles that drivers most often use to transport only themselves. As more Americans adopted this capsule model, US transportation infrastructure oriented itself around it via a proliferation of gas stations, rural roads, highways, urban streets, and suburban neighborhoods. These design choices, in turn, created new problems like traffic, long commutes, pollution, and high fuel prices.[46] Further, significant time, labor, and money are required of drivers to secure and maintain their ability to participate in the capsule model, not to mention that the demands cars place on us also function to isolate us from one another, as each driver is individually responsible for meeting the legal standards required to operate their car as well as for managing the economic burden of maintaining its mechanical functions. All of American society has been restructured around the individualism of capsule transportation. Rather than explore alternative models that might avoid problems caused by a proliferation of cars—such as intercity rail networks or publicly accessible, collective models of intracity transportation—Illich argues that American science and engineering are instead "organized to remedy minor inefficiencies that hold up the further growth" of social systems dependent on the capsule model. Each tweak or minor improvement to the infrastructure surrounding cars is thus "heralded as costly breakthroughs in the interest of further public service."[47] Even if the public were presented with alternatives, Illich concludes that they would refuse them. The spread of technologies built across the United States to support the capsule model would make any alternative seem too costly, cumbersome, unnatural, or otherwise inconvenient. Had they existed in 1973, Illich likely would have pointed to self-driving cars as nothing more than a new attempt by a rising group of technological elites to seize power rather than a genuine attempt to reform transportation.

According to Illich, the reason that most industrial technologies have hindered our pursuit of conviviality is because they are the product of institutions run by "self-certifying professional elites." On this topic, Illich uses the example of modern medicine. Technoscientific elites, he explains, have historically appealed to the idea of progress to justify their monopolization of skills and understandings that had previously been commonly accessible to communities to manage their own care, governance, and identity. All technoscientific revolutions thus follow a similar pattern. At their beginning, "new knowledge is applied to the solution of a clearly stated problem and scientific measuring sticks are applied to account for the new efficiency." The discovery of germs and the realization that simple hygiene routines could prevent common illnesses led to basic techniques that individuals could practice with minimal training. At some point, however, "progress demonstrated in a previous achievement is used as a rationale for the exploitation of society as a whole in the service of a value which is determined and constantly revised by an element of society, by one of its self-certifying professional elites." In other words, doctors eventually determined that the management of basic medical care could affect the outcomes of more complex treatments. Physicians thus began to insist that all medical care be professionalized. Basic knowledge and procedures were thus subsumed within an institutional framework dedicated to the pursuit of treatments for more advanced illnesses. Paradoxically, he concludes, a doctor's expertise is rewarded institutionally more by difficult patients who require greater medical care because their cases serve as an occasion for them to exercise the full array of the institution's resources (with each piece of their reasoning carefully recorded as a testament to their expertise).[48] Western scientific and industrial institutions are, in short, often structured so that the specialized literacies and technologies they produce privilege their elite members at the expense of the public they ostensibly serve.

"Tools," broadly defined, are often extensions of that privilege, lent to us as users for the purpose of reinforcing the authority of their designers.

By contrast, convivial tools are the product of institutions that foreground the values of survival, justice, and self-defined work. Illich cautions his readers that it is possible for institutions to prioritize each of these over the other. Prisons, for example, are in theory structured to ensure the survival of inmates so that they can serve out the full duration of their sentence, while denying them self-definition and prioritizing someone else's sense of justice. In order to support conviviality, Illich argues, we need to push for the public oversight of science and technology. Such oversight is the only way to guarantee that "no one person's ability to express him- or herself in work will require as a condition the enforced labor or the enforced learning or the enforced consumption of another."[49] The tools that these institutions produce must therefore be ones that "can be easily used, by anybody, as often as or as seldom as desired, for accomplishment of a purpose chosen by the user. The use of such tools by one person does not restrain another from using them equally."[50] The goal of scientific and technological institutions should therefore be to develop standards "in the public interest" rather than to maintain the self-certifying power of elite institutions or to press ever onward toward the profits of private interests. Tools should not be "owned" in the sense that any one group has a definitive say in their design or purpose.[51]

Illich's insistence that no one be able to monopolize technological power does resonate with Nelson and Levy. But, importantly, Illich's writings do not even implicitly call for the kind of technolibertarianism that today's Silicon Valley elites insist is necessary to ensure our digital freedoms.[52] Instead, Illich argues that the experience of true individual autonomy through technology does not depend on free access to tools but on the existence of socialized structures that check the power of elite actors. In other words, conviviality is not a product of the technology itself but of the society that technology is used within. Illich only mentions computers directly to emphasize this point and to remind readers of their potential to automate intellectual labor and thereby strip people of their autonomy. Computers could provide us with better access to information, but we should not "confuse vehicles for potential information with information itself. We do the same when we confused data for potential decision with decision itself."[53] Here, Illich explicitly cautions his readers to avoid the pursuit of technological expediency especially when it comes to computers: they could be used for self-improvement or in support of democratic movements, but they should not be mistaken for the source of either.

Contrary to Brand, Nelson, and Levy, Illich does not see technology as devoid of political valence. All technologies are the product of and participants in social relations. Technology is designed to support specific relations, and we must recognize that they are often designed in ways that make it difficult to separate them from those relations. This disagreement between Illich, on the one hand, and Brand, Nelson, and Levy, on the other, is key to understanding the continued popular acceptance of Silicon Valley's innovation mythology. The assumption that technology can be wholly divorced from the intentions of its designers helped to sustain a belief that personalness is possible even as computers began to recentralize around elite actors.

Revisiting Hobbyist Writings

Just as Levy's hacker ethic smooths over critical points of tension between the writings of Brand, Nelson, and Illich, so too does it impose a monolithic politics on the complex and conflicting attitudes represented in hobbyist newsletters. Accounts by Levy and others of early publications like the *People's Computer Company* and the *Homebrew Computer Club Newsletter* remain significant influences on American understandings of the politics of personal computing and the rhetorics of usability that support user-friendliness. However, popular histories like Levy's typically omit consideration of other prominent hobbyist publications like the *Mark-8 Users Group Newsletter* and *Computer Notes*. While there is some overlap in technological practice across all four of these newsletters, it is important to recognize that each had distinct goals and represented the idea of a personal computer revolution differently. Contributors to each situated themselves variously with respect to concerns about decentralization, personalization, and conviviality. As a result, each publication offers a different look into rhetorics and models of usability intended to realize a personal computer revolution. Together, they help illustrate the various ways that these groups responded when encountering opportunities for technological expediency.

Among the early hobbyist publications, the *People's Computer Company* was the most explicitly countercultural. While it does not map directly onto Levy's, Brand's, Nelson's, or Illich's politics, it does share some elements of each. One of its primary goals, which it shared with Nelson, was to foster widespread computer literacy and support the decentralization of computing. Yet the *People's Computer Company* did not advocate for the same highly individualistic models of usability described by Brand or Nelson. Instead, it called on readers to promote decentralization by establishing new independent educational clubs and local computing centers. Because it was published prior to the availability of individually ownable machines, the *People's Computer Company* assumed that its readers would be using either a networked, time-shared system or a cabinet-sized minicomputer. The *People's Computer Company* evoked a sense of conviviality by framing computing as a form of community engagement and emphasizing the importance of developing practices that would make computing seem inviting to anyone regardless of their prior technical skills. The *People's Computer Company* newsletter was part of a larger effort to realize a new approach to computing practice that also included the People's Computer Center, located in Menlo Park, and the Dymax Corporation, a publishing company that sold and distributed literature to assist with the acquisition and administration of shared computer hardware as well the design of curricula for its users.[54] What follows is a rough history of the newsletter's publication that devotes special attention to the way it negotiated the political frameworks associated with Brand, Nelson, and Illich. As I show, while the *People's Computer Company* shared some of Brand's and Nelson's concerns, it largely avoided the pursuit of technological expediency and instead foregrounded the sociocultural aspects of computing in order to promote a model of universal access that more closely, but not exactly, resembled Illich's conviviality.

The first issue of the *People's Computer Company* was published in October 1972. Initially, Bob Albrecht, the newsletter's founding editor, relied on the same equipment that Brand had used for assembling and printing the *Whole Earth Catalog*. Its inaugural issue roughly coincided with the end of Brand's initial phase of activism.[55] Even though Brand did not write for nor participate in the production of the newsletter, its visual style and page layouts during its first several years very much resemble the *Whole Earth Catalog*. Like Brand's publication, the *People's Computer Company* also included excerpts of material from other publications, mostly clippings from newspapers, computing journals, and engineering trade publications that discussed the cultural or social significance of computers. Alongside these, the *People's Computer Company* also printed the source code to BASIC programs, often with handwritten annotations inviting readers to experiment with altering certain lines or explaining the computational concepts present therein. Albrecht assembled the newsletter's early issues with help from Mary Jo Albrecht, Tom Albrecht, Jerry Brown, Marc LeBrun, LeRoy Finkel, and Jane Wood. Albrecht edited the newsletter until late 1976, passing control over to Phyllis M. Cole beginning with the January 1977 issue. The first twenty-six issues of the newsletter were printed as a black and white tabloid; however, starting with the May / June 1977 issue, the newsletter was reformatted as a magazine and renamed to *People's Computers*. Following the reformat, it dropped all nonoriginal content and reduced the amount of BASIC source code it printed. Regular columns discussed projects that readers could undertake with the first wave of home computers, and articles focused on creative uses that its staff or contributors had found for computers in the wild. With the July / August 1978 issue, the publication was once again rebranded as *Recreational Computing* and kept the title until it ceased publication in 1981. In its second and third iterations as magazines, the publication was one of the few early American computing magazines not to include commercial advertisements.

Although Albrecht and his collaborators make explicit political statements at several points in the first issue, this rhetoric becomes muted in later issues. The first issue features a masthead that reads "Computers are mostly used against people instead of for people, used to control people instead of to *free* them, time to change all that—we need a . . . People's Computer Company." Alongside the text is a drawing of a multiracial, intergenerational, mixed-gender group of demonstrators, seemingly protesting the oligopoly that large corporations and the government have hereto held over computing with signs that read "BASIC IS THE PEOPLE'S LANGUAGE" and "USE COMPUTERS *FOR* PEOPLE NOT AGAINST THEM!" While the images and language on the front cover of the first issue portray the publication's mission as one of reforming institutions through widespread computer literacy, the publication rarely engages with broader social and cultural issues directly. And although many accounts have pointed to, and continue to point to, the first issue's masthead as evidence of the publication's politics, subsequent issues use a revised masthead that does not include the drawing nor the calls for revolution. The drawing does reappear in a handful of later issues—usually as part of set of annotations in the margin of a source code printout buried deep within the issue—but never again as part of the masthead. The masthead in subsequent issues states that the *People's Computer Company* is "a newspaper about . . . having fun with computers, learning how to use computers, how to buy mini-computers, books, film, music, tools of the future." After the first few issues, explicitly politically focused writing was marginalized in favor of documenting software source code and providing other informational resources.

Even though overtly political language quickly disappeared from the publication, the *People's Computer Company* nevertheless did enact many of the practices described by Nelson. For example, the newsletter often appears to take up the call to challenge computer crud by providing plain descriptions of technical concepts. Each issue usually contained at least one article explaining some computational principle, offering examples in source code and inviting readers to explore different aspects of it by modifying the code at specific points. Unlike Nelson, however, Albrecht never outright states that a goal of the newsletter is to dispel common misbeliefs, and the newsletter never directly levels criticism at computer companies despite the occasional acknowledgment of the negative reputation IBM had among readers.[56] Nevertheless, Albrecht and his coeditors do frequently imply that demystifying computers would help minimize some negative social consequences of computing. The May 1973 issue, for example, includes a page with clippings that describe the *People's Computer Company*'s mission as showing people that computers "needn't be, and shouldn't be, objects of fear" by bringing "real, live interactive computing to the masses."[57] Even here, however, the *People's Computer Company* is careful to avoid suggesting that learning about computers was an individual's responsibility. The newsletter almost always represents its audience very explicitly and directly as educators who wanted to run programs similar to those that Albrecht offered at the People's Computer Center. In addition to sample code and lesson plan ideas, the *People's Computer Company* also commonly includes articles on topics like how to locate grants and other financing for minicomputers or examples of lessons and activities that clubs could run.[58] While the newsletter's efforts to establish spaces to explore the creative side of computing outside of academia and corporate America certainly had political implications, these articles are much more subdued when compared to Brand's calls to drop out of society or Nelson's characterization of his readers as computer revolutionaries.

Perhaps because they seemed like such a distant possibility, individually ownable computers are rarely discussed in the *People's Computer Company* prior to 1975. When they are mentioned, however, there is a tension present between the newsletter's broader gestures to something like Illich's conviviality and Nelson's digital freedom. An article from September 1973 by Gregory Yob, for example, describes the goal of building a hypothetical "brain for your boob tube" in language resembling Nelson's. He notes that owing to the high costs and complexity of current hardware, "access to a computer is usually through some kind of organization" and adds that "effective use of a general purpose machine [is] a highly technical skill." Yob laments that computer hardware "remain[s] inaccessible except to the specialists responsible for their maintenance . . . These tendencies clash with the 'dream' of a computer in every home." To address these problems, he sketches out an idea for "the design and construction of a computer for use by the general public without the need for any technical training." The computer system that Yob describes remarkably resembles the first home computers that would become available during the next decade. The computer is contained within a keyboard unit that uses cartridge-loaded software, a cassette recorder for storage, and a television as a display. However, the article itself is brief. He does not discuss its potential beyond offering a short list of vague applications like "auto tune-up, Form 1040, checkbook balance, calculation (business and scientific)[,] Teletype simulator, burglar alarm, automated house, [and] home studies courses."[59] Yob does not situate his article within the larger mission of the newsletter, which he could have accomplished by speculating about how his proposed personal computer might be incorporated into local clubs or what kinds of new social activities might be possible if every member had access to one.

At other moments, the newsletter represents personal computers as comfortably able to support conviviality. The December 1974 issue, for example, includes a lengthy essay by Lee Felsenstein that draws connections between his experience as a radio hobbyist, Illich's theory of convivial tools, and his hopes for computer kits. Felsenstein states his belief that the *People's Computer Company* had excelled in its presentation of software as a convivial tool and adds that his goal in contributing to the publication is to do the same for computer hardware. Felsenstein describes how he grew up tinkering with radios, noting that making simple repairs soon turned into building them from the ground up using spare parts. Citing Illich's discussions of preindustrialized technoscience, he explains that among radio hobbyists "people were building tools that other people could use without much training. Tools that people could use, and which would not use them. People could understand how the tools worked, how to fix them when they broke, and how to alter them when the job changed."[60] In other words, Felsenstein's experience had shown that it was possible for people to learn to work with electronics by experimenting, just as the *People's Computer Company* had been encouraging with its programming curriculum. Opposite his essay, Felsenstein includes a rough diagram of a hypothetical Tom Swift Terminal, showing how a computer with a modular model of usability would allow for the kinds of repairs and modifications he had made to radios. Despite going to great length to describe how the design of his computer would support individual autonomy and avoid creating an extractive relationship between users and designers, Felsenstein only briefly touches on the need to build communities and social structures around computers to maintain that autonomy. Radio hobbyism, he explains, had been set back significantly in recent years because radio manufacturers switched to using cheaper parts that were soldered to the board. These new radios could not be repaired or built from spare parts, only replaced. Felsenstein insists that there is no reason why computers have to follow a similar path and suggests that readers who feel the same should work with him to help build and manufacture his Tom Swift Terminal.

The next issue, published in January 1975 appears to announce the realization of Felsenstein's proposal. The front fold features a large image of the Altair 8800 computer, assembled from a kit manufactured and sold by Micro Instrumentation and Telemetry Systems (MITS). The newsletter also reprints on the same page a portion of the *Popular Electronics* article announcing the Altair 8800. The newsletter here presents the Altair 8800 as a solution to the financial problems that anyone looking to launch an educational computer center modeled after the People's Computer Center would face. Around the clipping, Albrecht and his coeditors add their own copy exclaiming that this is not just a computer you could buy "for your school" but was also one you could purchase "for yourself" or "your friend": "The home computer is here!" However, the text of the *Popular Electronics* clipping emphasizes radically individualized models of usability: the Altair 8800 is "a minicomputer that will grow with your needs, rather than one that will be obsoleted as you move deeply into computerized applications. . . . [Y]ou can be sure that there will be manifold uses we cannot even think of at this time."[61] Here, then, was when the *People's Computer Company* butted up against the possibility of an expedient solution to its mission to promote widespread computer literacy. Even though its engagement with American countercultural politics has been exaggerated by popular histories, the *People's Computer Company* repeatedly foregrounded the sociocultural aspects of computing, emphasizing them at least as much as the technical topics it covered. Initially, Albrecht resisted local contributors' suggestion that the focus of the newsletter be changed to home computer kits because he wanted to maintain a focus on community building and making a space for novices to explore computational concepts. When the Altair 8800 arrived, several of the newsletter's local contributors left to form a more elite group that would focus instead almost wholly on technical concerns related to the "dream" of a home computer.[62]

The next issue of the *People's Computer Company*, printed in March 1975, illustrates this split. Inside the issue were several full- or nearly full-page advertisements for newly formed groups that would be dedicated to supporting hobbyists who wanted to build their own personal microcomputers from kits. Readers in this issue could find information about joining the Homebrew Computer Club, the Mark-8 Users Group, and the Altair Users Group. The *People's Computer Company* notes that each would soon be or already was producing its own newsletter. For its part, the *People's Computer Company* would continue after the announcement of the Altair 8800 to focus on programming education with a particular emphasis on community building until mid-1977, when Albrecht stepped down as editor. His backing away roughly coincided with the first retail availability of more friendly, preassembled personal computers. Unlike the *People's Computer Company*, however, these three new publications had comparatively little interest in engaging anyone outside of their own subscribers or in cultivating community for any purpose other than building their personally owned machines. Petrick observes in her review of the *Homebrew Computer Club Newsletter*, for instance, that they only typically made efforts to share information when they were trying to complete a personal project they hoped to commercialize.[63] Any discussion of the principles or practices that Levy later codified in his hacker ethic are complicated during moments when a technologically expedient solution is presented.

Somewhat misleadingly, the Homebrew Computer Club is often characterized as extending the politics of the *People's Computer Company*. While there was initially some overlap between the two organizations, the countercultural commitments of the Homebrew Computer Club are overstated in popular histories, at least with respect to the club's self-representation in its newsletters. The *Homebrew Computer Club Newsletter* was originally edited by Fred Moore, the group's cofounder who was also a former instructor at the People's Computer Company center. Prior to his involvement with either group, Moore had been active in the antiwar movement, had been a reader of the *Whole Earth Catalog*, and had studied Illich's writings.[64] The first official issue of the *Homebrew Computer Club Newsletter* was published on March 15, 1975, and is little more than a single-page spare account of the first meeting. The second issue, however, established its format under Moore's editorship. Counting the first, Moore edited a total of five issues, departing from the group in August 1975. While it was loosely laid out in comparison to the *People's Computer Company*, the *Homebrew Computer Club Newsletter* did contain a similar mixture of hand drawn images and annotations as well as original typewritten content appearing alongside excerpts from other publications that were collaged together.

Despite adopting a sometimes flippant casual tone, Moore's descriptions of the group's activities are fairly banal. Perhaps the closest Moore comes to explicitly framing personal computing as countercultural is when, in the June 1975 issue, he explains his belief that the Altair 8800 would

> (1) force the awakening of the other [hardware manufacturing] companies to the demand for low-cost computers for use in the home, which will mean competition, resulting in lower prices just as happened with the hand held calculator, (2) cause local computer clubs and hobby groups to form to fill the technical knowledge vacuum, (3) help demystify computers. Computers are not magic. And it is important for the general public to understand the limits of these machines and that humans are responsible for the programming.[65]

These sentiments very much resemble those found in Nelson's *Computer Lib* with the addition of a brief, implicit reference to the *People's Computer Company*'s efforts to build communities around computers. Like Nelson, Moore here assumes that decentralizing computer literacy and enabling free access to computing power will allow the public to cut through the excuses that technocrats use to secure their power.

Felsenstein's voice is also prominent in early issues of the *Homebrew Computer Club Newsletter*. According to Moore, Felsenstein typically ran the group's loosely structured meetings, using his time as leader to encourage those in attendance to focus on community building. In the April 12, 1975 issue, for example, Moore summarizes remarks Felsenstein made during a meeting about how the club's newsletter could

> point to sources, items, news, etc. Sort of an identifier of people, places, articles, abstracts, and general information of interest to club members. It can serve as a link between members: who has what to share or who needs what. And that includes all of us. We each know something or have something—even if it is only time or energy. The assumption is we are all learners and doers. Right? The function of the club and the newsletter is to facilitate our access to each other and the micro-world out there.[66]

However, Moore's account does not make it clear how the club's members interpreted Felsenstein's remarks apart from suggesting that they recognized the technical advantages of cooperation with respect to improving the limited functionality of the computers they built from their kits. Nonetheless, from the second issue onward, the newsletter did include a listing of members. In addition to providing the name and mailing address of each member who asked to be included, the listings also offer brief statements regarding members' motivation for joining, a description of their skills, and notes about what kinds of hardware they need or help they are seeking for their personal hardware projects. Notable names among the listings are Felsenstein, John Draper, Jerry Lawson, Tom Pittman, Jim Warren, and Steve Wozniak.[67]

At other times, according to Moore's notes, Felsenstein explicitly advocated that the group pursue Illich's model of conviviality, pushing members to design systems that would be accessible to novice users. During one meeting, Felsenstein gave a brief lecture on his belief that the club should develop "a cheap terminal that would survive with untrained people" and called on others to "incorporate the user into the design" by "tak[ing] the place where the terminal is going to be installed and turn that into a little computer club. Then you've got your service organization right there, and they can get on very intimate terms with the equipment. And make it work their way. What this means is putting a sort of hobby center into each terminal. . . . This means building a device which can be expanded, and be modified, and is visible, is understandable, is convivial."[68] Unlike in his earlier contributions to the *People's Computer Company*, Felsenstein here makes a more direct effort to describe the kind of institution building needed to support the convivial machines he hoped to develop. He suggests that computer clubs could function as a source of public accountability for the computer industry, an accountability Illich felt was needed if it were to avoid the predatory designs common in Western industry at large. Computer clubs were well positioned between manufacturers and the general public because their members could use their advanced skills and knowledges to advocate on behalf of novices. However, Moore does not note any response to Felsenstein's remarks other than that "talk continued" after he had finished speaking.

While Felsenstein's discussion of Illich's ideas is one of the most explicitly political moments in the newsletter's run, it is also its last. Beginning with the August 20, 1975 issue, the *Homebrew Computer Club Newsletter* dropped almost all pretense of embracing counterculture and revolution. After Moore's departure in July, Robert Reiling assumed the role of editor, and from that point forward almost no discussion of technopolitics, implied, muted, or otherwise, appeared again. Reiling's accounts of the meetings more or less resemble formal minutes and are by comparison much more trim than Moore's, which tend to be wandering descriptions of specific moments that he found interesting during conversations among members. Reiling also shifted the format of the newsletter, making it much more technically oriented in nature. Readers would now find almost nothing but lengthy, jargon-laden proposals, circuit diagrams, and software in the form of instruction codes for the Intel 8080 processor. Although this kind of material did appear previously, the foregrounding of Moore's drawings and more conversationally styled writings at least suggested that anyone was welcome to join the club. Reiling's editorial style, by contrast, plainly indicated that both the Homebrew Computer Club and its newsletter were meant only for highly skilled enthusiasts. Thematically, the focus of discussion across the next several issues shifted to establishing technical standards for data storage, display drivers, and hardware interfaces, and the like. In this respect, the majority of the Homebrew Computer Club's newsletters and the activities recorded therein resemble those found in other similar publications that have been left out of histories of the personal computer revolution.

One such example is the *Mark-8 Users Group Newsletter*. In many ways, this newsletter established a model that the Homebrew Computer Club's newsletter would follow under Reiling. The Mark-8 was announced in July 1974, several months before the Altair 8800. Unlike in the case of the Altair 8800, the design and sale of Mark-8 kit was almost wholly decentralized. Designed by Jon Titus, the Mark-8 was "released" as a pamphlet. Titus had arranged for an electronics manufacturing company to print the circuit boards he described in his plans, but in theory interested hobbyists could arrange for their manufacture elsewhere. Other parts included commonly available electronics components and the Intel 8008 microprocessor, the one centrally manufactured component, which hobbyists could order directly from Intel.[69] Because there was no single manufacturer overseeing the design and implementation of the Mark-8, the *Mark-8 Users Group Newsletter* served as a primary resource for hobbyists to work out questions among themselves about standards beyond those minimal ones defined by Titus.[70] In short, Titus's design resembled the same distrust of technological institutions found in Brand's and Nelson's writings. It was a computer that could in theory be built and used in a context that was entirely removed from them. The newsletter was started by Hal Singer, and its first issue was published in August 1974. Unlike the Homebrew Computer Club, the Mark-8 Users Group had no in-person meetings, so the newsletter served as its primary forum for communal interaction. Apart from a handful of references to the "dream" of an affordable home computer, there is no revolutionary rhetoric. The newsletter only bears out the countercultural definition of personal computing insofar as it served, like Brand's *Whole Earth Catalog*, as a way for its disparate membership to share information and "tools" in the form of machine code and circuit diagrams. Like the *Homebrew Computer Club Newsletter*, the *Mark-8 Users Group Newsletter* included a membership listing with contact information in each issue. Many issues also featured summaries of the personal projects that members were working on as reported to Singer.

Although the newsletter does not directly consider the politics of personal computing, its members shared a radical distrust of centralization. In a handful of moments, Singer urges the group to address questions of standards that one can read as evocative of Illich's insistence that certain institutional frameworks were necessary to support a convivial technology. Without a centralized manufacturing source, the members of the Mark-8

Users Group were in a position to develop their own technological standards as they refined Titus's original design. Early issues contained detailed proposals for members for various peripheral, circuitry, and software standards, but Singer laments at several points that Mark-8 users were never able to agree on which standards the community should adopt, preventing the Mark-8 from becoming a platform that readers could use to share data and software with one another. Reviewing the results of a survey of reader activity, for example, he comments in the April 1975 issue that "everyone's development work has headed in a different direction and our only hope now is to hop onto someone's bandwagon that has done important development work." He explains that many of the problems that the Mark-8 Users Group was struggling with were ones that did not exist for Altair 8800 users, who could rely on MITS to settle these kinds of questions. Singer then points to two small companies who were planning to sell add-on kits that would introduce many of the features its readers could not come into agreement over and suggests it would be better to adopt their products than continue on as they had been. His comments here suggest that the Mark-8 failed to gain the same traction as the Altair 8800 because its users were unwilling to compromise in their pursuit of decentralized design. Of course, neither of the manufacturers that Singer recommended resembled the sort of publicly accountable institutions that Illich argues is necessary to ensure a technology's conviviality, but their products would have provided the group with a technologically expedient solution to their unending disagreements over standards. The Mark-8 thus did not become popular for the same reason that Brand's communes did not succeed. The revolution could only advance if there were some form of governance over the technology, someone who could make decisions that would transform the members individual Mark-8 kits into a shared platform. The kind of highly individualized personalness that Brand and Nelson called for could not here realize a viable model of usability.

Commercializing the Hobbyists' Revolution

While it is not clear whether the struggles of the Mark-8 Users Group were widely known, there is some evidence that other hobbyists were concerned about uncertainties over standards. This concern can be seen in the way other groups rejected early efforts to build core components convivially in favor of adopting proprietary versions simply because the latter were available first. Anticipating the proliferation of kits based on Intel's microprocessors, Dennis Allison proposed to readers in the March 1975 issue of the *People's Computer Company* the idea that they work together to "Build Your Own BASIC." Small computers, he explained, would need a dialect of the language that was suitable for the reduced amount of memory they would have available.[71] Readers were intrigued and wrote in to express their support and offer their assistance in developing a "Tiny BASIC." Allison soon followed up with a more detailed description of how to implement the language.[72] In other words, Allison and Albrecht supported the development of a convivial technology: software designed and managed a by an organization that made its decision-making transparent, invited open participation in the design process, and acted on input from the community it served. However, many hobbyists had already begun to adopt a commercial BASIC interpreter developed by Microsoft and sold through MITS. Even though Tiny BASIC interpreters were available for essentially the cost of shipping the paper tape they were printed on, they did not arrive until mid-1976, almost a year after MITS began selling Microsoft's interpreter.[73] Looking more closely at the rhetoric of MITS's Altair Users Group, we can begin to see how a rhetoric of usability circulating among hobbyists was revised to accept the use of privatized, centrally managed technologies in the interest of technological expediency.

Initially, the Altair Users Group and the *Computer Notes* newsletter that MITS oversaw espoused similar ideas to those of the earlier groups and publications I have discussed. For example, while MITS did not print source code in *Computer Notes*, it did provide free and low-cost software to its users through its Altair Library program. To build the library, MITS solicited submissions from readers as part of a "software contest." The contest was overseen by Bill Gates, who worked as a contractor for MITS during this time. As the first announcement for the contest explains, members of the Altair Users Group were encouraged to "submit programs for the *Altair Library*. . . . All programs will be screened and tested by MITS. Once a program has been found to be acceptable, it will be included in the *Altair Library* and a description of the program will be printed in the User's Club newsletter. The author of the program will be entitled to a free printout of any programs from the *Altair Library*."[74] No prices for software in the library are noted; however, later issues indicate that library software was available to members for a small copying fee, anywhere from $2 to $15. Newly accepted software and award winners were listed in each issue, with the library expanding by a dozen or more programs each month. Although initially submissions were only in machine code, Gates noted in July 1975 that they had begun receiving a small number of programs written in Altair BASIC and that he expected many in the future as more and more users setup their computers to support the language.[75] Although perhaps a bit more cumbersome than the *People's Computer Company*'s approach in the sense that readers would have to request copies of specific programs and pay for access, MITS's software contest does indicate that an early goal of *Computer Notes* was to act a kind of public-facing organization through which Altair users could participate in defining the usability of their computers.

Regardless of however genuinely committed its staff may have been to the idea of community engagement with its hardware, the newsletter also acted as a catalog for MITS's products. Altair BASIC was a key item in this respect. Editors and contributors to *Computer Notes* were not shy about framing Altair BASIC as a critical contribution to the personal computer revolution. The front page of the very first issue, for example, declares that "there are two keys to the new computer revolution. One is computers must be inexpensive and the other is computers must be understandable. With the Altair 8800 and Altair BASIC, both of these criteria have been met."[76] While Altair presented the software as its own, Altair BASIC had been produced by Gates, his business partner Paul Allen, and a contractor named Monte Davidoff. MITS licensed the software for resale from Gates and Allen's company, then known as "Micro-Soft." Initially, MITS listed Altair BASIC at an á la carte cost of $500 and the more advanced Altair EXTENDED BASIC at $750. MITS also charged a $30 copying fee for each. These costs could be reduced, however, if users purchased either version of BASIC as part of a package deal that waived the copying fee, bundled at $75 and $150, respectively.[77] Notably, prices for both versions of BASIC when purchased individually were more than the hardware itself, initially listed at $439 for the entry-level kit.[78]

By October 1975, MITS had evidently received enough complaints about the cost of Altair BASIC that the company's president, Edward Roberts, felt the need to respond to them in *Computer Notes*. Observing that a significant number of users had suggested that "MITS should give BASIC to its customers," Roberts responded by explaining that "MITS essentially makes no profit on BASIC" due to its royalty commitment to Microsoft. Quality software is expensive to produce, and users should recognize the service that MITS is providing the community by "selling the BASIC at 1/10 to 1/100 the price large computer companies would get for a similar package."[79] That same month, MITS relisted its BASIC and EXTENDED BASIC packages for $200 and $350 each, offering the previous discounts of $75 and $150 to anyone who had previously purchased a kit and the upgrades necessary to support the interpreters.[80] Despite trying to cultivate a reputation for software sharing through its library program, piracy of Altair BASIC was a recurrent theme during the first two years of *Computer Notes*. After the software first became available in July 1975, for example, *Computer Notes* printed reminders in subsequent issues that Altair BASIC would not be sent to customers unless they first returned a signed software agreement which specified that they would not use the software on any computer other their own Altair 8800.[81] Additionally, the newsletter's editor, Dave Bunnell, frequently echoed Roberts's remarks in his editorials, explaining that MITS had been performing an important service for its users by making BASIC readily available to them.

Gates took the arguments further, suggesting that he and Allen had solved a critical technical problem by developing their BASIC interpreter. In February 1976, the newsletter printed an "Open Letter to Hobbyists," signed by Gates, which alleged that "as the majority of hobbyists must be aware, most of you steal your software. Hardware must be paid for, but software is something to share. Who cares if the people who worked on it get paid?"[82] As Kevin Driscoll notes, the histories that discuss the letter, including Levy's, tend to frame the conflict around it as evidence that the hobbyists' countercultural ideals had led them to resist Gates's attempts at commercializing software. Although Gates derided hobbyists, his point that without financial incentive, quality, bug-free software like Altair BASIC is not possible seemed reasonable to many of them. Driscoll notes that responses to the letter published in various newsletters and early computing magazines were largely sympathetic to Gates's argument even if they bristled at his criticism of hobbyists. Many also praised the quality of Microsoft's work and even joined Gates in shaming the alleged pirates. Other respondents suggested that those hobbyists other than themselves who were pirating BASIC were likely only making copies of it to protest MITS's prices rather than out of any political commitment to free sharing. The high price was essentially locking people out of personal computing, they complained. Even those hobbyists who were openly critical of Gates's insistence that BASIC be paid for directly by users rather than be free or sold at a much lower cost nonetheless conceded that quality software like Altair BASIC was essential to personal computing's future.[83]

The hobbyists' preference for Altair BASIC did not constitute a turn away from countercultural politics but was rather a continuation of their privileging of technological concerns over the cultural, social, and political issues emerging around computing. The hobbyist responses to Gates's letter printed in other newsletters and early magazines generally only disagreed on whether or not the cost of BASIC should be built into price of their kits.[84] They changed their rhetoric of usability to justify the adoption of a high-quality, privately managed component, believing that its widespread adoption would improve the usability of the Altair 8800 for all users. Gates contributed further to this shift in a follow-up letter published in the April 1976 issue of *Computer Notes* by arguing that personal computing would not be possible without elite users capable of providing tools and established standards that would help make these new machines usable and useful for more than just individual tinkering: "Software makes the difference between a computer being a fascinating educational tool for years and an exciting enigma for a few months and then gathering dust in the closet." If users wanted to be able to share their own software with one another, that was their business. But they if wanted tools to build it, then it was their responsibility to ensure that Microsoft or any other organization with the resources needed to develop quality tools received "a reasonable return on the huge investment in time that is necessary."[85] Gates's second letter is an early example of Silicon Valley's innovation mythology, as it frames his company as uniquely capable of supporting some aspect of the personal computer revolution. Our popular image of Silicon Valley's techno-elites is certainly influenced by American hagiographies of industrial-era inventors. However, by the 1980s, the popular press was also keen to promote the idea of elite computing professionals who were able to succeed in business while appearing to bring elements of counterculture into the board room.[86]

A Revolution for the Rest of Us

The 1977 West Coast Computer Faire introduced a similarly revised rhetoric of usability that positioned elite designers as capable of realizing the revolution not just for other hobbyists but also for the general public. Here, a focus on technological expediency moves more firmly toward an embrace of universal usability while negotiating lingering expectations for digital freedom. A similar negotiation can be found in Apple's advertisements following the event. Popular histories of the fair tend to portray it as the first major hobby computing event, drawing thirteen thousand attendees.[87] Yet this impression is misleading in a number of ways. While large, it not the first major event to include home computers in the United States. The Trenton Computer Festival, for example, held its first meeting a year prior, using a format similar to the one that the West Coast Computer Faire would adopt.[88] The January 1977 Consumer Electronics Show also included among its exhibitors the Commodore PET, a computer designed for use by the general public rather than hobbyists.[89] Additionally, popular histories misleadingly imply that most participants were hobbyists. For example, Levy describes the event as "wall-to-wall technofreaks" and a "hardware hacker's version of Woodstock."[90] However, reporting from the show floor suggests that a significant number of people outside of the hobbyist subculture were present. Both Ted Nelson and Mike Markkula, a co-owner of Apple then serving as the company's vice president of marketing, are both quoted as being surprised to see so many novices in attendance. As Nelson comments when asked about his experience there, the fair showed that the "computer world has suddenly been broke in two. There's the straights and the strange coalition of hobbyists."[91] And when asked if Apple was prepared to sell its new Apple II computers to the novices in attendance, Markkula enthusiastically says yes and promises that the machine was capable of providing every user with the fruits of the hobbyists' personal computer revolution right out of the box. Users would be able to play games, use productive software, and even write their own programs within minutes.[92] If Gates framed the revolution as one that could only be realized for hobbyists if core technologies were managed by elites, the fair represented it as something that elites would provide for everyone else.

As staunchly opposed to the private control of information relating to technology as Nelson was, even he was beginning to concede that the revolution would need to be managed by a new group of elites. At the fair, Nelson delivered a speech in which he declared that he saw "little reason to sell to hobbyists, and considerable reason not to."[93] From a financial standpoint, the market was limited. Also, hobbyists complained too much about prices. These concerns aside, however, the future of computing could not follow a hobbyist model of usability if a social revolution were to be realized. The goal for "highly interactive [computer] systems will be making things clear and simple. Now, making things clear and simple and easy to us is, I'm afraid, the opposite of what some computer people, perhaps some of you, want to do. So in an important sense, it is *not* the laymen who have to learn Computerese: in the greater sense, WE MUST ALL LEARN COMPUTER EASE!"[94] The hobbyist model would soon "correspond to the tin-can and crystal era of radio; and the convivial hobby you are part of right now may vanish like the crowd that welcomed Lindberg at Orly. They don't come out to meet the planes anymore. Today's summer-camp camaraderie won't last forever, and the computer will probably become a home appliance, as glamorous as a canopener [sic]."[95] Mass computational literacy, in short, should no longer be our primary goal. Instead, Nelson argues, entrepreneurial hobbyists were beginning to shift their design efforts towards questions of universal usability and fairgoers would be wise to turn their attention to novices, too.

This was not the first time Nelson had suggested that hobbyists would soon assume a priesthood-like role, shepherding society through a new information age powered by personal computers. In a speech delivered at the World Altair Computer Convention the previous year, Nelson claims that "it's the people playing with the switches in their garages who are going to make computers as practical for the household and small businessmen as they are for massive corporations." After stoking up revolutionary fervor by invoking an image of personal computing in opposition to IBM, Nelson shifts his tone abruptly, "predict[ing] that the phenomenal growth rate of hobbyists will level off. Why? 'Because not many people like to program or wield a soldering iron.'" The emphasis on hardware hacking that had animated early participation in personal computing was preventing the revolution from moving forward. Now, a model of universal usability would be necessary: "Computers are going to recapitulate the history of hi-fidelity." That is, "the household computer will be as common as the . . . turntable" and "the technological knowledge of the computer will be exclusive among owners as hi-fidelity know-how is today." Nelson, however, states outright that universal usability does not mean a universally shared computer literacy. In fact, tomorrow's computers would be managed by today's hobbyists: "There will always be computer insiders-like us, and computers outsiders-like them, but that's all right." It really isn't necessary, he continues, "to feel that we have to impose on the world the learning of ASCII code."[96] To be fair, Nelson does not say that this new elite of hobbyist entrepreneurs should hoard knowledge, simply that not everyone should be required to master computing to benefit from the technology. At the same time, his position here represents a significant change from his earlier insistence on mass computer literacy as a political imperative. You no longer can and must learn about computers now. You still can, if you want, but not if its inconvenient. As long as the new, reformed priesthood would benevolently make its innovations available to us all, it did not matter as much if everyone learned all of the technical concepts involved.

Apple's earliest advertisements further illustrate the way that ideas associated with hobbyists' discussions of digital freedom were reframed to fit the new focus on universal usability. Although its Apple I single-board computer sold poorly, the advertisements Apple designed for it would establish a framework of hobbyist ideas that it would repurpose later for novice users when promoting the Apple II. Apple's first advertisement was published in the July 1976 issue of *Interface Age*, a magazine associated with the Southern California Computer Society. Like MITS, Apple touted the importance of its proprietary BASIC yet implied that it was holding truer to the hobbyist ideal of shared access to tools when it announced triumphantly that "Yes, folks. Apple Basic is free!" The second advertisement for the Apple I, which began appearing in October 1976, establishes the Apple I as more than a kit. It was, instead, a "truly complete microcomputer system" that "opens many new possibilities for users and systems manufacturers." Like MITS, Apple also presented itself as trying to engage with its community of users, stating that the company promised to "provide software for our machines free or at minimal cost" so that buyers "won't be continually paying for access to this growing software library."[97] In these early advertisements, Apple is signaling that both the company and its computers are products of the personal computer revolution. Not only did the company support the goal of providing free access but it would also act as a convivial steward by assisting its users in locating tools and information to advance their use.

While the Apple I was clearly a machine intended primarily for advanced hobbyist users, the Apple II was one designed to serve as a link between hobbyists' desire for freedom and the universal usability needed to appeal to novices. Even though Apple advertised the system as "turnkey," by 1978 it had released a complete technical reference and an extensive handbook for programming in both the Apple II's machine code and the proprietary Apple BASIC dialect.[98] All of the knowledge that early groups had acquired, whether they shared it openly or not, was now made available with the machine. The handbook assisted third parties who wanted to develop hardware expansions or develop professional quality software themselves, which helped sustain consumer interest in the machine for almost a decade. Apple also tried to make tools freely available to its users through the Software Bank library program that it arranged in partnership with retailers. For a small fee that they paid to a partner store, Apple II users could copy to tape or disk any or all of the software applications included in the source disks that the company sent to its partners. Like MITS did for its Altair Library, Apple solicited submissions from its users and packaged together ones that had demonstrated "usefulness" and were the "exemplification of good programming practice."[99] Especially in the early life of the system, before the software market around it had had a chance to grow, these programs would have provided a sense of immediate utility, offering far more than the one or two pieces of demonstration software that most computers shipped with at the time. It is not clear how much this distillation of hobbyist values embedded in the Apple II's design appealed to novices, however. Most accounts of the company's history agree that the machine sold well primarily due to its appeal among business users and through Apple's concerted efforts to sell the machine to schools.[100]

Nonetheless, the Apple II's rhetoric of usability linked freedom and universalness through an explicit focus on novice users. The Apple II took the revolution's countercultural values and turned them into a package that could comfortably fit in the home or office. The first advertisement, for example, features a bold header that reads "You've just run out of excuses for not owning a personal computer."[101] The advertisement's copy alludes to many of the ideas expressed in the hobbyist community: the Apple II is "infinitely flexible," suggesting that users would ultimately determine its purpose, and a tool for developing computational literacy through experimentation that will "ready [users] for an evening of discovery in the new world of personal computers" as they learn to "write simple programs . . . or invent [their] own games." Like most of the advertisements Apple would produce over the next decade, it also assures readers that the machine would "grow with you as your skill and experience with computers grow." Instead of a bare circuit board, intimidating front display panel interface, or other overtly technical images common among other hardware advertisements appearing in computer magazines at the time, the Apple II is shown sitting on the kitchen table. Although many of the practices that the Apple II was designed to support were similar to those valued in hobbyist newsletters, the machine nevertheless invited the unexperienced, novice user to take part in a different kind of revolution: one that is exciting and yet still safe and familiar. The future of computing could be explored by anyone right from the comfort of their living room.

Apple's advertisements from then on would increasingly emphasize the utility of its products. Although there was always a sense of exploration or creative repurposing in advertisements for the Apple II, by the 1980s most mention of technical details would be gone, replaced by promises of how it would change readers lives at home, work, or school.[102] If creativity, free access, or the ability of users to shape the computers to suit their own needs were mentioned, it was almost always in the context of productivity. The revolution had arrived, and it was no longer just something that a handful of hobbyists benefited from. Apple's innovative hackers turned engineers had emerged out of that subculture, its advertisements implied, harnessing the hobbyists' values and sharing them with the rest of the world. While Levy and others began to draft their accounts of the personal computer revolution in the early 1980s, Apple would be preparing to advance its rhetoric of usability even further, promising that its next computer would truly bring a revolution "for the rest of us." A focus on expediency would continue to play an important role in the discourse around personal computing, and a preference for comparatively simpler, technical solutions would repeatedly be pursued in response to concerns voiced about the complex cultural, social, and political problems associated with computing.

Conclusion

The idea that the personal computer revolution must be privately managed on our behalf to support a model of universal usability has today reached a new apogee in the form of cloud computing.[103] Over the last decade or so, we have become acclimated to automatic updates and digital purchases, but now the software we are interacting with is increasingly managed elsewhere. In addition to email, messaging, and productivity applications, even software as demanding as media editors and video games can now be executed in the cloud and streamed to our phones, laptops, or gaming consoles. Cloud computing has taken us from the radically decentralized early visions of wholly individualized personal computers back to the te minals used before the era of personal computing to log into mainframes and time-shared systems. Beyond simply pulling computation back to centralized sites, cloud computing also grants technologists complete control over use practices. Cloud software cannot be pirated or modified. Our access can be granted or denied by toggling a flag on our accounts. Glitches can be removed instantly, making exploits or cheats a thing of the past. And yet when we look at how cloud computing is discussed in popular discourse, it is represented almost uniformly as strengthening our digital autonomy by making computing more convenient, more accessible, and cheaper than the alternatives. It obviates the need for users to purchase expensive hardware upgrades or to wrangle with troubleshooting their devices. If companies like Apple, Google, and Microsoft—who provide and manage the infrastructural hardware and software we rely on to make use of our personal computing devices—were to declare future support only for cloud computing, what option would be left to us but to go along or else cut ourselves off from digital culture?

Revisiting the sources of the personal computer revolution nearly four decades later makes it clear how closely shifts in the rhetorics and models of usability that support digital culture bear out Illich's cautionary tale about medical science. Before 1975, the *People's Computer Company* presented computing as a technology that communities could tailor and manage to meet the needs of their local users. The development of microcomputer kits was hailed as a great breakthrough, but it quickly necessitated the expert management of even its most basic components in order to support higher-level uses. Along the way, hobbyists held up the importance of individual autonomy but moved further and further away from a convivial idea of computing as the need to sustain software companies became paramount. Following the years covered in this chapter, Microsoft's BASIC would become the de facto standard in personal computing. Most of the manufacturers discussed in the next two chapters either licensed existing versions from Microsoft or hired it to write a new version specific to their machines. In either case, most personal computing activities in the United States began to be supported by Microsoft's code. The number of manufacturers of core components like operating systems, processors, and wireless chipsets has continued to shrink, giving a small handful of companies significant power over our digital media ecology and forcing users to pay for their mistakes.[104] Further, the interplay between digital freedom and universal usability continues to position technologically expedient solutions as preferable to any kind of public oversight. Consider, for example, the recent growing concerns over misinformation and harassment online. Rather than develop, implement, and refine widespread server-side controls, Big Tech has responded with rhetorics and models of usability that emphasize the individual responsibility of users. It is our responsibility, they tell us, to block what we do not want to see. Conveniently, they fail to note that it is also cheaper and easier to outsource that responsibility to users. Indeed, Big Tech companies only move to correct themselves in the face of public outcry, when they feel the need to protect their finances. It has become clear, as Illich predicted, that computer technologies require a form of public accountability that does not hinge on popular reputation, stock value, or perceived utility.

Today, the major computing companies in the United States style themselves as the inheritors of the revolution's values. Despite dropping almost all pretense of being countercultural, giants like Amazon, Apple, Facebook, Google, and Microsoft still promote the idea that their technologies are uniquely designed to empower us to smile, think differently, connect with one another, do good, and fulfill our potential. They present themselves as the alternative to less imaginative areas of American industry and finance. Many of the specific principles of Levy's hacker ethic were left by the wayside as computing commercialized, but the idea of hacking as a skill that sets today's innovative elite apart from the rest of society remains. Indeed, as Lilly Irani has observed, elite technologists who view themselves as hackers often regard themselves as better able to understand the needs of the users they design for than the users themselves.[105] Despite hacking's incorporation into today's techno status quo, activists, critics, and academic researchers who call on us to push back against the politics of Big Tech still position hacking as a form of resistance. Like Illich, I remain skeptical that a technology can be wholly divorced from the politics of its creators, especially when it exists within infrastructures that enforce those politics, although the recognition that all technology is socially constructed and thus can be appropriated for new purposes unanticipated by their designers is still valuable. But given the way political principles associated with hacking historically have helped both to justify the pursuit of technological expediency and to downplay the sociocultural costs of expediency—and, in doing so, have provided a foundation for the embrace of hacking as innovation in contemporary corporate techoculture—it is time that we reconsidered the language of our resistance. Practices that we label as "hacking" and that are intended to explore undervalued aspects of computing technology can too readily be folded into Silicon Valley's favored rhetorics of usability as acts of resistance are deftly parried and reframed as entrepreneurial activity.

To some extent, the embrace of the practices and products of the free and open source software (F / OSS) movements within American tech culture illustrates the way that appeals to principles in the hacker ethic allow resistance to be reframed in support of Silicon Valley's pursuit of technological expediency. Advocates of these movements have implicitly styled themselves after Levy's hackers, promoting a new, revolutionary technological philosophy centered on the information sharing practices of Brand and Nelson and suggesting that their politically transparent design and management of software will help us realize a more democratic society.[106] But it is important to recognize that F / OSS groups and projects are not free from the influence of the cultural and social norms they claim to resist. Indeed, the fact that F / OSS technologies are being used within and are contributed to by programmers working for Big Tech companies is less of an issue than that F / OSS's rhetoric of usability similarly serves to obfuscate its privileging of technology expediency over the cultural, social, and political concerns its movements were ostensibly formed to engage with. During its first two decades, F / OSS participants viewed meritocracy as a simple solution to worries about social organization: the best coders, they assumed, would naturally be recognized as project leaders. Yet in the face of a growing number of incidents of identity-based harassment, programmers like Coraline Ada Ehmke developed—and in many cases successfully lobbied for the adoption of—"codes of conduct" that would serve both as formal commitments to inclusion while also offering projects enforcement mechanisms to handle future harassment.[107] Many committed F / OSS participants objected, claiming that if marginalized groups felt they were being treated poorly then they simply needed to learn to write better code. Frameworks like Ehmke's have thus helped to expose not only the ways that F / OSS's narrow focus on technical solutions has allowed discrimination to go unchecked but also how some F / OSS luminaries were able to launder far-right views through the movements' democratic trappings.[108]

If nothing else, this chapter should serve as a reminder that we must approach the rhetorics of usability promoted by a technology's designers with suspicion. There is no simple approach to resisting Big Tech's computing power because its power is complex in its structure and its effects. Technologists want us to see ourselves as participating in a revolution because it helps to draw our attention away from the motivations hidden within their designs and toward fantasies of convenience, connection, and self-improvement. In the next chapter, I discuss how even though many of the companies making the first retail-friendly "appliance computers" were eager to disassociate themselves from hobbyists and counterculture, they nonetheless worked carefully to convey the idea that a revolution was happening as a way of persuading consumers that the efforts they took to seize control of early computing ecosystems were ultimately in the public's best interest. Long before today's walled garden appstores, the rhetorics of usability connected with the appliance computers of the 1980s pushed users to associate the aggressive actions that hardware manufacturers took to control third-party software developers as the best way to achieve user-friendliness.

3

Appliance Computing

The Revolution Comes Home

In American popular culture, the personal computer revolution of the late 1970s and early 1980s is often portrayed as a heady time when home computers sold faster than manufacturers could fill orders. But many observers writing during that period offered a more sobering account. Wayne Green, for example, was an early computing journalist who was very enthusiastic about the future of the industry. In addition to founding over half a dozen computing magazines, he also owned a software publishing company. Writing in 1980, Green commented that while he expected the industry would "grow significantly for business uses," he was "not wholly convinced of the place of the computer in the home, or even the concept of the personal computer. I suspect that the media have been led astray by these terms."[1]

As I discussed in the previous chapter, the early years of personal computing are remembered primarily through a countercultural lens; however, if we look away from the hobbyist-focused accounts that invited people to imagine an intellectual exploration of computing from the comfort of their living room, there was another, much more conservative narrative of revolution circulating within American culture that represented computers as essential to the future of business. These newer, more affordable computers had the potential to give small businesses and individual knowledge workers access to tools that had previously only been available to larger firms or giant corporations. Several early personal computers that are downplayed in popular histories despite outselling the Apple II for several years, like the Tandy TRS-80 and Commodore PET, were developed and promoted with an eye to this more conservative narrative.[2] Returning to the conversations taking place about them can help us to understand how the rhetorics of usability associated with these less well-documented "appliance computers" negotiated these two contested visions of a personal computer revolution in the United States. As I argue in this chapter, it is through this negotiation that early articulations of user-friendliness began to take shape prior to the emergence of a consensus around transparent design.

Even when appliance computers are discussed in popular histories of personal computing, they are often mocked by their authors. This derision likely owes to efforts by their manufacturers to explicitly distance their products from hobby computing. The TRS-80, for example, is often remembered as the "Trash 80" and has been dismissed in popular histories of computing as the technological equivalent of fast food: cheap, unsatisfying, and poorly made.[3] The PET is similarly remembered more for its calculator-style keyboard than for any other contribution it made to American personal computing.[4] These machines may not have introduced any lasting standards; however, they did serve as a point of first contact for millions of American who wanted to experience the "personal computer revolution" that they had read about in the news. For better or worse, in other words, machines like the Tandy TRS-80 and the Commodore PET were important influences on early conversations taking place across the United States during the late 1970s and early 1980s about what it meant for personal computers to be usable and useful to the average person. In the pages that follow, I show how these two machines, along with computers that had a similar design, like the Atari 800 and the Texas Instruments TI-99/4A, introduced the concepts of appliance computing and user-friendliness to American consumers.

Before proceeding further, I want to emphasize that "appliance computing" remains a significant influence on our contemporary computing practices. Lori Emerson characterizes appliance computers as machines that were meant to be used with only a "perfunctory know-how," just as with "any home appliance." Suggesting that Apple's Lisa and Macintosh established the norms of appliance computing in the mid 1980s, Emerson observes that, ever since the Macintosh was introduced, Apple has distinguished its products from those of its competitors by reframing drawbacks like a lack of expandability as a major convenience. Today, the company continues to promote a model of appliance computing through its phones, tablets, and App-Store.[5] The next major push for appliance computing, she notes, came from Mark Weiser, who throughout the 1990s described a model of "ubiquitous computing" that called on designers to "weave [computers] into the fabric of everyday life until they are indistinguishable from it."[6] While Emerson focuses on tablets and other mobile devices as examples of the way computing has been integrated into everyday life, Paul Dourish and Genevieve Bell note that discussions of appliance computing after Weiser also have proposed embedding computers in other mechanical and electrical technologies.[7] Whether we are talking about smart phones or smart homes, the idea that we should be able to treat computers as just another kind of domestic appliance remains popular.

Although it is true that Apple's products and Weiser's writings have strongly influenced the design of today's handheld, mobile, and "smart" technologies, the idea of appliance computing can be traced back further to at least 1977. The Apple II, Tandy TRS-80, and Commodore PET were all initially described as "appliance computers" because they promised users "a complete system presented in an assembled and tested package."[8] But computer magazines and mainstream publications alike soon began to group Tandy's and Commodore's computers separately from Apple's, citing the latter's support for expansion and much higher price point. Tandy's and Commodore's computers were at various points also referred to as "small computers" and "home computers," indicating that they were generally not seen as entry-level business machines but as computers for playing games and other nonserious uses. Soon, machines similar to the TRS-80 and PET like the Atari 800 and Texas Instruments TI-99 / 4A entered the market. By the early 1980s, "appliance computing" had become a term that was used to contrast these lower cost, black-box machines from the "general purpose computers" that were understood to be adaptable to a wider variety of professional purposes. In short, appliance computing prior to the Macintosh typically referred to a class of computers that intentionally compromised expandability and computing power in order to support a new model of immediate and universal usability.

Aside from allowing us to understand better the social construction of user-friendliness, examining the rhetorics and models of usability associated with the appliance computers of the late 1970s and early 1980s can also help us to consider the politics of today's "walled garden" app stores. A walled garden software ecology is one that a platform designer is able to exert a very fine degree of control over by explicitly requiring that each application run on it go through a formal approval and permissions process. So far, only

Apple's AppStore has implemented a true walled garden, but Google's and Microsoft's stores are taking steps in that direction. Despite placing severe constraints on both third-party software developers and users, the App-Store has earned high praise from technologists due to the way it is understood to provide a simple user experience for locating and installing software.[9] As I show, the first steps toward forging an association between user-friendliness and the aggressive management of software ecologies were taken by manufacturers of early appliance computers. Whereas the Apple II's success has been attributed, at least in part, to its open design and well-documented hardware standards, Tandy, Commodore, Atari, and Texas Instruments each in different ways and to varying degrees attempted to limit or outright block unauthorized third-party hardware and software development.

In describing their rhetorics and models of usability, I am attentive to the way that early appliance computers functioned as "platforms," which are the "underlying systems" of computer technologies that "enable, constrain, shape, and support the creative work that is done on them."[10] Specifically, like Tarleton Gillespie, I understand that platforms have politics insofar as they allow their designers to exercise power over computational culture. While Gillespie considers this power in connection with YouTube, I look at it in relation to appliance computers, detailing how their manufacturers implemented designs in response to financial and cultural demands that allowed them to appear to consumers as neutral intermediaries supporting a growing, general interest in computing while at the same time influencing the shape of the media ecologies that were supported by their computers.[11] Notably, Tandy, Commodore, Atari, and Texas Instruments developed reputations among advanced users as being at the very least reluctant to share technical information publicly. Some not only shared technical information about their products sparingly but also even took steps to exert control over computer shops that sold their products, used warranties to discourage users from opening their machines, and in certain cases threatened software developers who did not sign distribution agreements with legal action. At the same time, engineers, executives, and other company spokespersons worked to persuade consumers that their black-box approach would make computers more usable and useful.

In sum, this chapter documents how an apparent concern for "ease of use" can be, and often is, appealed to in order to justify other purposes that are often not openly or fairly represented to users. I begin by reviewing the more conservative vision of a personal computing revolution advanced outside of specialized computing publications in American print media. Compared to popular narratives of this period, many journalistic accounts portrayed the personal computing revolution as one focused more on business uses, and even when they did turn their attention to home use, they typically indicated there were fewer good reasons for owning one. Early reporting by journalists without prior experience in computing also portrayed personal computers as unnecessarily difficult to use, and many were often unsure of the benefits they might provide to the average person. The rhetoric of appliance computing would thus map a focus on immediate and universal usability from business use onto home computing. Next, I consider what appliance computing meant in terms of system design by examining how manufacturers and journalists established a separate category of preassembled, retail-friendly machines that they assured consumers would be immediately usable and useful. Finally, I examine the specific rhetorics of usability circulating around the Tandy TRS-80, the Commodore PET, the Atari 800, and the Texas Instruments TI-99/4A in order to document the circumstances of their design, how they framed their designs as responses to perceived public attitudes about computing, and how they leveraged the idea of user-friendliness to exert control over third parties that contributed to the retail environments in which their machines were being sold.

A Different View of the Personal Computer Revolution

While a countercultural vision of the personal computer revolution has remained the focus of many popular books about digital culture in the United States, it is important to recognize that this vision was not well represented in the American mainstream press until the mid-1980s. As I discussed in Chapter 2, books by technology journalists like Steven Levy helped to establish a durable narrative that associates personal computing with American counterculture and that continues to privilege technological expediency in our understanding of user-friendliness.[12] Yet, as Thomas Haigh notes, it was business computing that was the focus of much of the early reporting on microcomputers in national newspapers and magazines during this same period. The popular press was "awash with discussion of new, or newly popular, buzzards such as microelectronic revolution, the information society, the chief information officer, information technology, the postindustrial society, and knowledge workers."[13] While Haigh's account of the popular discussion of computing centers on expensive office automation technology, it is not hard to find examples of similar rhetoric in portrayals of microcomputers in offices and factories. Even as articles asked readers to imagine what it might be like to have a computer in their home, they often went into much more detail about the advantages that new "personal computers" would afford small businesses who had previously not been able to justify the high cost of office automation systems manufactured by companies like IBM, DEC, or Wang Laboratories.[14]

News articles discussing the future of business computing explicitly referred to personal computers as more "friendly" than mainframe or minicomputers. Computers had already been "shrinking," and many who followed the industry assumed that it was only a matter of time until major manufacturers moved on from cabinet-sized minicomputers to desktop-sized microcomputers just as they had moved on from mainframes previously. Rather than hobbyists or hackers, many mainstream journalists in the mid-to-late 1970s looked to IBM and its competitors to describe the future of small computers. A June 1977 article from the *Washington Post*, for example, included statements from executives at Honeywell, IBM, and UNIVAC describing how small computers were being put to use in a variety of productive contexts like factories, farms, and hospitals. Home use was mentioned, too, but executives still treated it as a distant possibility. While articles like this one suggested to readers that it was much easier to imagine new uses for computers in corporate and industrial contexts, they also brought up ideas that would become important in later discussions of user-friendliness. As one executive explained, these new computers would be "personalized; they're not Big Brothers. They're friendly machines[,] . . . they're your machines." They also stressed that these new, smaller machines represented "the first beginnings of the utter involvement of the layman in computer technology."[15] Here, they leveraged the ideas of friendliness, personalness, and universal usability to describe how readily these computers could be put to a variety of productive uses outside of specialized laboratories: testing complex machines as they come down the assembly line, managing livestock populations, and keeping track of patient medical histories. To be sure, there are some distant similarities between the optimism expressed in these articles and those bold promises made by early hobbyist visionaries like Ted Nelson in that they both imply that the decentralization made possible by personal computers would lead to endless possibilities for innovation. Generally, however, their excitement reads as much more fervent when discussing work-related activities than home or leisure use.

There were also stories that described the personal computer revolution as having potential negative impacts on American society. But like those that offered more uniformly positive visions, they too focused more on business and industrial contexts. Often, moments of doubt occurred in stories that were otherwise utopian and futurist; reporters would briefly wonder whether certain portions of the population would be displaced as computers reduced the demand for human labor or eliminated certain professions entirely. In 1976, for example, a feature in *U.S. News & World Report* promised an "explosion" of uses to come:

> Hardly a facet of American life will be left untouched in the age of the computer now taking shape. The changes the computer has made already in American life are insignificant compared with the startling advances predicted for the coming decade. Every corner of society is going to feel the computer's impact. . . . Businesses, large and small, will use computers increasingly to make key decisions, not just to keep records. At the same time, though, there are grave social consequences that seem to prevent computer technology from being applied willy-nilly.[16]

The potential concerns, the article noted, were being voiced by unions, consumer groups, privacy advocates, and teachers. Some feared that the pursuit of increased profits were pushing companies to integrate computers into their offices and factories without really considering how the automation, the need for retraining, or the isolation associated with computer use would affect workers. The article also stated that the coming computer revolution would affect households primarily via the changes to retail and banking. Use in the home, however, would likely only become common after increased commercial usage had helped to bring down the cost of computers.

When journalists did try to imagine how the personal computer revolution would change life at home in more concrete terms, they often described a future that left the American status quo largely intact. Compared to articles discussing business uses, they were also much more vaguely utopian.[17] *Time* magazine's "The Computer Moves In" issue, for example, suggested that

> the enduring American love affair with the automobile and the television set are now being transformed into a giddy passion for the personal computer. This passion is partly fad, partly a sense of how life could be made better, partly a gigantic sales campaign. Above all, it is the end result of a technological revolution that has been in the making for four decades and is now, quite literally, hitting home.

Citing a survey conducted by *Time*, the article noted that "nearly 80% of Americans expect that in the fairly near future, home computers will be as commonplace as televisions or dishwashers."[18] When trying to describe what this new appliance would do for its users, however, the article was more circumspect, explaining that "there are many people who may quite reasonably decide that they can get along very nicely without a computer." As an appliance, it offered few compelling uses: "Why should a computer be needed to balance a checkbook or to turn off the living-room lights? Or to recommend a dinner menu, particularly when it can consider (as did a $34 item called the Pizza Program) ice cream as an appetizer?"[19] Perhaps most representative of a comparative lack of imagination when it came to discussing home use was the fact that most articles pointed to the same, uninspiring list of activities for use in the home: personal banking, filing taxes, indexing recipes, planning trips, managing a calendar, and potentially accessing a news service if you were willing to pay extra for a data connection.

Cost-benefit discussions of computing figured regularly in articles about personal computing in mainstream news publications. Many articles were written with at least a partial focus on helping readers decide whether or not to purchase a machine for themselves. For example, in a series of articles for the *New York Times* documenting his own experiences as a novice computer user, journalist Peter Schuyten regularly observed that the only real reason to buy a computer for home use was to satisfy one's own curiosity. With respect to practical uses, they seemed to hardly live up to the promises he had heard about: "The central question about the market for these systems seems to be whether the consumer really wants to pay $500 to $3,000 to balance the checkbook. At this point, the answer would seem be no."[20] When chronicling his own purchasing experience, he explained that he looked forward to teaching himself to program but admitted that not all of his readers would find it appealing.[21] Buyers guides, too, began to appear in a wide variety of magazines. Generally, they all ensured readers that a personal computer would not feel out of place in the home, but they often compared them to toys or exercise equipment, suggesting that they might be fun to try or provide some modest form of self-enrichment but were far from essential. Whether they were worth buying at all, many suggested, depended on whether or not you believed all the talk about how they would change the working world. If you did, then the most compelling reason to buy one would be so that your children would not be left behind.[22] As I discuss in the next chapter, this fear of potentially being left behind in an increasingly computerized society was part of a growing concern over a computer literacy crisis that exerted a strong influence on the conceptions of user-friendliness that took shape as appliance computers fell out of favor.

Hobbyist Computing vs. Appliance Computing

While many of the articles discussed in the previous section were published after appliance computers were already gaining popularity, the concerns they expressed were ones that companies like Tandy, Commodore, Atari, and Texas Instruments developed rhetorics of usability to position their products against. During the late 1970s, the business utility of computing was generally acknowledged and accepted; however, there was a broadly shared skepticism regarding the practical value of personal computers, especially compared to more expensive, office automation systems. Companies like Tandy and Atari tried to emphasize that their computers were value-priced options for serious purposes even as the media coverage of them generally assumed that their low-cost machines were more appropriate for gaming and leisure. By comparison, Commodore and Texas Instruments more explicitly pitched their computers as designed for home use either by comparing them to domestic appliances or by selling their products in department stores. Regardless, manufacturers of appliance computers generally all emphasized that their products had clear uses and provided real value that anyone could realize with minimal effort. In short, the rhetorics of usability associated with appliance computing leveraged ideas from business computing in order to transform home computing from a technology with a limitless but ultimately abstract potential into a retail-friendly product designed for specific, practical purposes.

When discussing appliance computers, American journalists tended to categorize differently those computers primarily available at specialized shops and those available at general consumer electronics or department stores.[23] Appliance computer manufacturers wanted their users to have a fairly mundane retail experience, one that did not subject them to a consideration of a litany of technical details and that potentially even encouraged impulse buying. Descriptions of computer retail environments published in the late 1970s and early 1980s typically depicted specialized computer shops as difficult spaces for computing novices to navigate. Shops that opened during the mid-1970s often functioned as "overgrown clubs" that were started to provide a local group of users with discounts on parts through bulk purchases. Because customers and staff alike were hobbyists, it was not uncommon for people to bring their personal machines into the shop, where they could tinker with them and also consult others for advice on their projects.[24] Shops like these were often not welcoming to first time buyers, either because their employees were not trained to explain technical concepts to novices or because they became visibly annoyed when asked basic questions.[25]

Specialized shops that catered more to novices had different problems. Many were managed by people with backgrounds in sales who were drawn to computing as a business but knew very little about computers themselves. Stores managed under these conditions might have a sales staff trained to give misleading advice, policies that encouraged customers to buy expensive machines that exceeded their stated needs, or a service department that incorrectly diagnosed issues to charge customers for unnecessary repairs.[26] Buyers guides generally warned first-time buyers to come in prepared and offered them scripts to follow when visiting specialized shops.[27] While appliance computers were often sold in these types of stores during the late 1970s and early 1980s, unlike the Apple II they were also sold through familiar, friendlier retailers like Radio Shack, K-Mart, and J. C. Penny.[28] While the emergence of customer-service oriented, national chains like ComputerLand would erode this distinction somewhat, there were other ways that consumers were encouraged to view appliance computers as part of a separate class of machine.

American journalists covering the personal computing industry also typically divided their coverage along a $1,000 line, labeling those at or below it as "small" or "home" computers. Whereas Apple began selling its basic Apple II model at $1,298, Tandy first advertised its basic TRS-80 at a cost of $599.95.[29] The Commodore PET was first sold in the United States for $800.[30] During the 1980s, journalists would also include Texas Instrument's

TI-99 / 4A (initially $525 and later $99) and Atari's 400 and 800 ($550 and $1000, respectively) in this class of computers.[31] In truth, these machines were not much cheaper than the kit-based computers available at the time. Companies manufacturing kits based on the Altair-standard like Cromemco, IMSAI, and MITS advertised preassembled machines at a cost of $1000, $1100, and $680, respectively.[32] More expensive options were available from all of these companies, of course, but if price were a primary concern then there were already low-cost options available. What set these newer machines apart from the cheaper kits was their presentation as appliances that could be setup within minutes and a black-box design that promised users would never have to engage with hardware directly.

Before turning to the conversations unfolding around each machine more closely, I want to take a moment to consider what this black-box approach to design meant in practice. To today's readers, the idea of a closed architecture machine might not seem unusual, especially if they do most of their computing on a laptop, own a mobile device, or play video games using a console. In the late 1970s and early 1980s, however, most computing magazines treated open designs that allowed users to access internal components as the norm, especially for computers above the $1,000 line. This open design typically featured an internal "expansion bus" with slots for circuit board "cards" that could add new functions to the system. Companies like Apple and IBM supported the development of internal expansion by including detailed technical references with their computers. By comparison, appliance computers were designed with relatively fixed internal configurations and did not include much documentation apart from demonstration software and short "getting started" guides that walked users through turning them on for the first time. Some also included an introduction to BASIC programming, and most manufacturers would eventually provide BASIC programming manuals tailored to the specific dialect of their interpreters. However, complete manuals were not available for early adopters and later were not always included in the price of the machine. Today's devices similarly have exteriors that are designed to either prevent them from being opened at all or at least to prevent them from being opened easily. Small upgrades like battery replacement are still possible on some of them, but most users do not install internal expansion cards, with the one common exception being gamers who choose to install high-end, dedicated graphics processing cards into their desktop computers. Instead, most expansion today occurs externally through USB peripherals. Features like Bluetooth connectivity, wireless networking, increased storage, access to obsolete storage formats like floppy disks or optical media, and video capture can all be added through self-contained external devices or dongles. The appliance computers of the late 1970s and early 1980s were similar in their focus on expansion through external devices, but each executed their black-box design slightly differently.

The TRS-80 itself resembled a bulky, plastic keyboard and came with a monitor and external power supply unit.[33] The monitor was a small black and white television that did not include a speaker or a tuner and was modified to display a command-line interface at a sixteen by sixty-four character resolution. Initially, the TRS-80 only supported storage via cassette tape. Tandy sold separately a cassette recorder that it claimed was designed to work specifically with the TRS-80. However, any cassette device could work with the machine so long as it was set to the right volume and speed. Tandy would later develop a floppy disk system for the TRS-80, but using it required the purchase of an expansion interface. The expansion interface was a boxy external peripheral that could easily be mistaken for the computer itself. Most photographs in Tandy's advertisements show the expansion interface positioned beneath the monitor and behind the keyboard. Although the expansion interface could in theory support other uses, for most users it was only configured to house the controller card needed to access the disk drive. Somewhat confusingly, the TRS-80 supported two different versions of the BASIC programming language. Initially, the TRS-80 was sold with Level 1 BASIC, which supported only a limited version of the programming language. In 1978, Tandy released an upgraded version called Level 2 BASIC that replaced Level 1 BASIC with an interpreter developed by Microsoft. This upgrade was one of the few internal expansion options available, but few TRS-80 computers were typically sold without it after its release due to the popularity of the disk drive. Users could also purchase more RAM for their TRS-80s. However, Tandy's warranty forbid users from opening the keyboard unit and instead required that the few internal upgrades available be performed by Radio Shack technicians. Service charges for these upgrades were often expensive, and bootleg do-it-yourself kits eventually appeared in some independent shops.

Unlike Tandy's TRS-80, the Commodore PET had most of its components integrated into a single unit.[34] A monitor, keyboard, and cassette tape storage device were built into the case, which could be raised via a rear hinge similar to that used to prop open a car hood. Somewhat awkwardly, because the keyboard, monitor, and cassette storage device were built into the top of the machine, opening the case also meant lifting all of these components together. Like the TRS-80, the PET did not support internal expansion. As Commodore noted in its promotional material, this easy form of access was meant for appliance service professionals rather than consumers. Comparing it to a television set, Commodore explained that users were not expected nor encouraged to perform any maintenance themselves and should instead take it to the same shops they patronized for repairs to their other appliances. Assuming that users would be more familiar with calculator keypads than full-size computer keyboards, the original PET used a "chiclet" style keyboard with a compact layout. The PET only supported expansion through a single external port. As with the TRS-80, PET users who wanted to use a disk drive needed a manufacturer installed upgrade to support BASIC 3.0.[35] While the upgrade provided the commands for users to access the disk drives, Commodore's drives contained their own control electronics and firmware that gave the PET access to a disk operating system when connected. The external disk drive units were almost as large as the PET itself.

Marking the beginning of a second wave of appliance computers, Atari's 400 and 800 were announced in late 1978 and available for order in early 1979.[36] Like the TRS-80, both the Atari 400 and 800 were keyboard units designed to be connected to a television set for use as a display. However, Atari did not sell dedicated monitors for either machine. While the Atari 400 and 800 both supported cassette tape storage, an external disk drive unit was available at launch and could be used without any upgrades or additional supporting peripherals. Similar to the TRS-80 and PET, both machines had only a single external peripheral port; however, Atari's port was designed so that additional external units could be "daisy-chained" together to allow for multiple expanded functions to be accessible at the same time. Additionally, the Atari 400 and 800 also had a small plastic lid located above the keyboard that could be opened, providing controlled access to system internals. Both machines supported ROM software cartridges that were modeled after ones used on Atari's video game consoles and that could be loaded into the system via a slot beneath the small lid. While application software was available on cartridges, Atari also used the internal cartridge slot for its BASIC interpreter, which meant that users could purchase and install upgrades themselves by replacing the cartridge that came with the system. While the Atari 400 only had a single cartridge slot, the Atari 800 featured a second, allowing for specialized cartridges that extended the functionality of others, as well as four memory expansion slots. By the mid-1980s, due to popular demand, the Atari 800 was generally only sold with all four memory slots fully upgraded. Unique among early American appliance computers, Atari tried to create machines that incorporated elements of internal expansion while limiting access to most of the system.

Texas Instruments was one of the last major American companies to introduce a new appliance computer standard prior to the release of the IBM PC.[37] The TI-99/4A, released in 1981, was a revision of the TI-99/4, released in 1979. Like Tandy's and Atari's early computers, the TI-99/4 and TI-99/4A were self-contained keyboard units that were designed to use televisions as displays. Texas Instruments did manufacture a dedicated monitor that was available separately for $450.[38] While there are some significant differences internally between the TI-99/4 and TI-99/4A, they were similar externally except for their keyboards. The earlier model was built with chiclet-style keys, whereas the TI-99/4A included full-size keys. Both of Texas Instruments' machines used ROM software cartridges. Texas Instruments' cartridge slot was more prominently visible than Atari's, located to the right of the keyboard and not hidden by a flap. The TI-99/4 supported only cassette tape storage, but an external disk drive peripheral was available for the TI-99/4A. As with the TRS-80, a separate expansion interface unit called the TI-99/4A Peripheral Expansion Box (PEB) was available for the computer; however, it was not required to use the TI-99/4A's disk drive. The TI-99/4A's PEB was larger than the computer itself. When opened, its internals resembled the bus of a general purpose computer, allowing users to slot in full-size expansion cards. The most common cards increased memory or added support for new external connection types. The PEB also included a built-in disk drive and supported connections for two additional drives. Separate, single-function peripherals were also available and could be daisy-chained together like Atari's.

Although each of these black-box designs was presented to the public as a way to make computing more accessible, they all were strongly influenced by a combination of internal factors unique to each company and the shared external influence of the more conservative vision of a personal computer revolution popular in the American news media. These companies all made explicit moves to limit the ability of users to experiment and develop new, third-party products as a part of their business strategy. A number of them even said the quiet part out loud. An executive at Texas Instruments, for example, commented in 1982 that the company expected consumers to spend as much as $3 on software and peripherals per year for every $1 they spent originally on their computer.[39] Commodore's CEO, Jack Tramiel, similarly told journalists on several occasions that he believed that "depending on outside sources for major components was the wrong way to run a business." He often bragged about how the company's vertical manufacturing allowed him to finely control the price of the PET and its peripherals.[40] The reviews and articles I examine in the next section indicate that all four companies took an aggressive stance toward hobbyists looking to develop hardware and software for appliance computers. To varying degrees, each company withheld technical information, pressured third-party developers into strict publishing contracts, or attempted to exercise control over the retail environment. While observers who participated in or were sympathetic to the hobbyist subculture—including many technology journalists—often viewed elements of appliance computing as anticonsumer, these four companies nevertheless developed rhetorics of usability that reframed their business practices as part of a commitment to user-friendly design in a way that many observers responded favorably to. The constrained models of usability they developed, they assured the public, were intended to alleviate concerns that computers were not usable by or useful to the average person.

Tandy TRS-80

While accounts of Tandy's development of the TRS-80 disagree on some details, nearly all indicate that the executives overseeing its production were either dimly aware of hobbyist computing or outright dismissive of it.[41] Tandy was certainly familiar with hobby electronics. In addition to selling electronic appliances like televisions, Tandy's Radio Shack stores also sold kits and parts for radios. Each individual transaction was small, but a significant portion of Tandy's business strategy was oriented around cultivating repeat customers by encouraging them to tinker continually with their personal projects. Yet Tandy's executives felt that computers would not appeal to the same hobbyists who were interested in the personalness of their radios. Due to its much higher cost, the TRS-80 needed to effect a sense of immediate utility. Lewis Kornfeld, president of Radio Shack during the period of the TRS-80's development, claimed that "there were no known [Radio Shack] customers" asking for computer kits. He noted when executives first began discussing the possibility of manufacturing computers that "it was virtually impossible to identify buyers" through their stores.[42] However, he and other executives believed that businessmen would be comfortable with a price that was higher than that of Radio Shack's other appliances if the company's computer were recognized as a serious productivity tool: "I'm not a computer anything. But my rules were: No kit, we wanted something that worked out of the box. No funny names, you know, like apples or oranges or lemons. As I put it, no racing stripes. We wanted it to look like a piece of business equipment."[43] Tandy maintained this focus on business utility throughout most of the TRS-80's lifespan. In 1981, for example, Tandy's then director of computer merchandising, Ed Juge, commented that the company's "primary [interest] has always been the business user," and the company "only advertise[d] to homeowners during Christmas time."[44] Among the companies pushing an appliance model of computing in the late 1980s, Tandy stands out as the most focused on situating its products as part of a personal computer revolution framed around business uses. Tandy would later introduce appliance computers explicitly marketed toward home use, like the TRS-80 Color Computer, but they were generally presented as intended primarily for gaming, positioned as toys within the expanded line of later TRS-80 variants.

Given that Tandy wanted to dissociate the TRS-80 from hobby computing, it is somewhat surprising that the company ended up hiring a West Coast hobbyist like Steven Leininger to lead the design of the TRS-80. Leininger has stated in interviews that he had regularly attended meetings of the Homebrew Computer Club and had seen Wozniak's early demonstrations of the Apple I prototype.[45] According to David Ahl, Leininger had also been involved with the People's Computer Center in Menlo Park, helping to adapt Tiny BASIC to the minicomputer the center used for its programming classes.[46] Leininger has also commented in interviews that he subscribed to several computing publications like *Creative Computing*, a magazine that very much tried to effect a sense of counterculture by mimicking the *People's Computer Company* newsletter in its artwork, layout, and prose styles.[47] By day, Leininger was an engineer at National Semiconductor, but he also worked part time some evenings and weekends at the Byte Shop Computer Store because it gave him a way to participate in hobby computing. The store was one of several listed as a recommended shop in the *Homebrew Computer Club Newsletter*.[48] Yet, as Ron White notes, Leininger was more "conservative" than other members of the club because he more or less embodied the status quo of the engineering industry: he was a clean-cut, college-educated, married, white male with a full-time job at a large company.[49] Several accounts of the TRS-80's development suggest that Leininger was offered a job with Tandy after he commented to executives who were visiting to source parts from National Semiconductor that public interest in computing had remained limited primarily because "too many people don't know how to solder."[50] Although many kit manufacturers and computer shops were happy to assemble machines for those buyers willing to pay extra, Leininger's remarks suggest that he viewed personal computing similarly to Tandy's executives. The decentralized model of usability favored by hobbyists and supported by computer kits was not one that would persuade the general public of the usableness and usefulness of personal computers.

Tandy's initial advertisements for the TRS-80 featured a rhetoric of usability directed toward nontechnical users, promising potential buyers that the machine would be both affordable and immediately useful. The initial 1977 advertisement for the TRS-80, for example, showed the machine being used for various tasks at home on the kitchen counter, in the office, and in the classroom. The advertisement's copy noted that the basic package included "everything you need to start using it immediately" and claimed that i' could be used for "personal finances, small business accounting, teaching functions, kitchen computations, [and] innumerable games."[51] This same list of uses would reappear throughout Tandy's first few years of advertisements; however, only business uses would see any real degree of elaboration. The next series of advertisements that Tandy ran throughout 1978 featured a table that listed each of the standard TRS-80 packages sold through Radio Shack and its price along with a picture of the machine with included peripherals as well as a brief description of the package's contents. The only uses implied by the advertisement were signaled by the more expensive package names: "Breakthru" ($599), "Sweet 16" ($899), "Educator" ($1,198), "Professional" ($2,385), and "Business" ($3,874).[52] Beginning in the 1980s, Tandy's advertisements for the TRS-80's successors would become more verbose and descriptive, but for the first model, the suggestion to readers was simply that it was a low-cost ready-to-use-out-of-the-box machine built for business but also able to support some productivity or leisure-related activities in the home.

Tandy's Radio Shack catalogs provided a larger space for the company to explain to potential buyers how and why personal computers might benefit them. Here, the difference in the level of detail in the description of professional versus home uses is much more noticeable. The 1978 and 1979 editions, for example, began by assuring readers that the TRS-80 is ideal for "the small business, laboratory, classroom, and the home." The early pages also included a section titled "What does a computer do?" that framed the machine as built for business but potentially adaptable to other purposes:

> Large computers are well known in the business world for their ability to do bookkeeping, billing, payroll, inventory control, and analysis and forecasting business data. Laboratories, engineering firms and universities have used computers to analyze volumes of data and numbers in a wide variety of applications. . . . The TRS-80 Microcomputer System is capable of performing all of these activities. In general, it differs from its larger cousins in speed and the amount of information which can be kept on line at any one time.[53]

While the catalogs also showed photos of the TRS-80 in the home and in the classroom, they did little to help clarify how the machine might be used in those contexts. Business uses, on the other hand, received considerable attention. For example, the 1979 catalog included a full-page table listing office activities in one column and detailing which packages or peripherals were necessary to accomplish them in the other columns. In short, Tandy's advertisements offered a stark contrast to the way personal computing was represented in hobbyist circles. Unlike Apple in its early advertisements, Tandy made no attempt to associate itself with hobbyist culture. The TRS-80 was not presented as a tool for exploring the technical side of computing from the comfort of your living room. Instead, Tandy's rhetoric of usability emphasized that the TRS-80 would be immediately usable by everyone and able to support a specific set of activities that users would already be familiar with.

Many hobbyists responded unfavorably to Tandy's design and marketing by arguing that its appliance model of usability was taking advantage of the fact that most people did not know enough about computers to evaluate them properly. For example, a review for *People's Computers* argued that the TRS-80's "cheap electronics" were a poor introduction to personal computing:

> The entire feel of the system with its display and keyboard encased in light plastic is more like that of a toy than that of a seriously designed computer and certainly not like that of a business machine. The TRS-80 represents, in my view, an attempt to capture a vast consumer market that is ignorant of the details with a quick and cheap machine and is disservice to the personal computing industry as a whole.[54]

Among the review's complaints are the "kluge" of three power cables required to operate the keyboard-unit, monitor, and cassette recorder; the unreliability of saving and loading from cassettes; limitations of the included BASIC interpreter; and the poor quality of documentation for the buggy software included with the machine. Many readers, however, evidently felt differently, and a later issue included several letters from users defending the TRS-80's construction. Several noted the reviewer had a strong bias toward the Commodore PET and suggested that the objections were trivial once users upgraded to Level 2 BASIC.[55] For its part, Tandy did seem aware of the TRS-80's poor reputation among some hobbyists, but the company seemed to assume that this view was not widely shared among people new to computing. Kornfeld, for instance, once commented that "the hobbyist, while vocal and visible, is not the mainstream of the business."[56]

Some hobbyists, however, did praise the TRS-80's design, suggesting that its appliance style would serve as a good platform for the commercial software industry. Software publisher Dan Fylstra, for example, explained in his review that the closed architecture made the machine "as foolproof as possible." As an "appliance," it "brings the personal computer a good deal closer to the average consumer." However, Fylstra had some doubts about Tandy's promise of immediate and universal usability, noting that the demonstration software appears to fall short. Nonetheless, he expressed confidence that in the long term Tandy will succeed in its goal of "support[ing] primarily . . . the small business market, although applications for education, entertainment and home use are clearly contemplated." He further suggested that the machine would likely come to be seen as more valuable over time as "knowledgeable hobbyists and small systems developers" began to develop software to fulfill Tandy's promises about its utility in the small business market.[57] Before working for Tandy, Juge reached similar conclusions in his review written for *Kilobaud Computing*. Like Fylstra, Juge suggested that the TRS-80's appliance design made computing far more accessible than earlier machines:

> Until now, I have never seen a hobby computer you could carry home from the store, plug in and use in anything other than machine language (numeric codes). The manufacturers apparently have been so deeply into computers themselves that they failed to realize how many people out there in the real world have no experience or training with computers, aren't interested in constructing a unit or having to program with lights, switches or even numeric codes.[58]

While hobbyist revolution narratives promoted a sense of personal freedom by suggesting that people should be able to creatively explore computing with few constraints, Juge defended the TRS-80 appliance design as one that is still within the "true computer-hobby spirit" because it made computing power immediately accessible to a wider range of users.[59] For Juge, Tandy's emphasis on practical applications was simply a better way to communicate the benefits of personal computing to a nontechnical audience. For those willing to learn to program, Juge concluded, there's nothing about the TRS-80's design that would prevent them from freely exploring potential uses for the machine. The readily accessible BASIC interpreter would allow anyone to develop their own software "for almost anything. Just use your imagination."[60] Even Bob Albrecht of the *People's Computer Company* claimed that it was likely the best computer he had seen for introducing new users to BASIC. Even though other machines, like the Apple II, may have been more impressive on a technical level, he felt that the TRS-80's simple design and low price were ultimately more important when it came to introducing people to computing.[61]

Although some hobbyists saw the TRS-80 as a promising software publishing platform, Tandy limited sales of the machine to its Radio Shack stores and did not permit stores to stock anything other than Tandy-branded hardware and software or products from third parties it had established partnerships with. According to Green, by 1980 there was a large community of programmers producing software for the TRS-80. Yet

> Radio Shack can't advertise this because they are trying as hard as they can to keep this fact a secret from their customers. They don't want TRS-80 buyers to know that there is anything more than a handful of mediocre programs available. These are the programs put out by Radio Shack and sold from their stores. Some might call this greed, others might term it practical business sense. That depends on whether you are working for Radio Shack or independently. . . . [Tandy-]owned stores are not allowed to sell any products not made by [Tandy] . . . [and cannot] even let the customers know that such exist. No books or magazines which hint at outside sources are permitted to be sold in the stores.[62]

Green was likely in a better position than most to observe this behavior. His software publishing company, Instant Software, primarily sold programs for the TRS-80, and one of his magazines, *80 Microcomputing*, was largely dedicated to it. In most of his columns, Green generally defended most of Tandy's business practices. However, Green consistently voiced concern over Tandy's treatment of independent software developers and independent retailers, arguing that its attitude toward third parties would threaten the TRS-80's perceived utility among the business users it claimed to prioritize. These policies did not stop third-parties from supporting the TRS-80, but Green suggested that they made the user experience more complicated than it needed to be. Consumers might find themselves in a store that sold TRS-80 software but no computers or not be able to find a piece of software they had read about because Radio Shack refused to carry it.

Compared to other appliance computers, the TRS-80 stands out due to Tandy's ability to control the retail environment in which it was sold. Tandy's approach—its insisting that the TRS-80 only be sold through its Radio Shack stores and its dictating which hardware and software could be sold through those stores—is reminiscent of the walled garden strategy of today's app stores. Tandy consistently tried to frame the dependence it fostered among users on Radio Shack as a way of prioritizing technical service in ways its competitors could not. At least initially, users responded favorably, with one remarking that taking a TRS-80 to be serviced "wasn't even inconvenient, especially with local Radio Shacks all over the place."[63] Despite Green's concerns, retailer Stan Veit observed that other appliance computers "never came close to the TRS-80 in popularity and customer satisfaction." In those areas of the United States where specialized computer shops could not gain purchase, the TRS-80 became a mainstay of schools and small businesses.[64] Importantly, the TRS-80's enclosed hardware also helped to introduce a rhetoric of usability that defined user-friendliness as achievable only through a model of usability that limited access to certain aspects of computing. Although reviews may have criticized the machine's cheap construction, its black-box design did not cause much concern, nor did the fact that the TRS-80's warranty required that users take it to Radio Shack for servicing and upgrades. These design features seemed to be quietly accepted as the way the personal computer revolution had to be reframed if computers were to better address the needs of the novice user.

The Commodore PET

Although the original model of the PET did not sell well in the United States, Commodore's early promotion of it developed a rhetoric of usability that like Tandy's associated a closed design with immediate and universal usability. Because the PET was the first, preassembled, non-kit-based machine that many computing magazines were able to obtain, early accounts of it took up the question of how the "appliance computing" model it offered fit within the then accepted norms of personal computing. Additionally, remarks by Commodore engineers and executives in hobbyist magazines about the PET introduced many ideas about how best to address usability for novices that would be carried forward by other companies into the early- and mid-1980s. In contrast to Tandy's rhetoric of usability, however, Commodore's more explicitly associated the PET with home use, often making explicit comparisons between the PET and other domestic appliances. In making these comparisons, Commodore challenged the hobbyists' ideal of personalized computing by promoting the idea that these machines should be viewed primarily as sources of personal convenience rather than as tools for intellectual self-empowerment.

The Commodore PET began as a project devised by engineers at MOS Technology as a way to sell more of its 6502 processors.[65] The MOS 6502's lead designer was Chuck Peddle, a career engineer who had little personal interest in hobbyist computing. After manufacture of the MOS 6502 began in earnest, Peddle started to travel the country to advise businesses about how they might incorporate the processor into their products or services. As part of this work, Peddle and other engineers at MOS developed a pair of single-board computer kits that would serve both as demonstration units for the 6502 and as a tool for engineers to learn to program the 6502. MOS eventually offered one of the kits for sale to the public, and it quickly became popular among hobbyists. The KIM-1 was a preassembled, single-board system that allowed its users to start programming out of the box; however, users did have to supply a terminal or teletype interface if they wanted to do more than enter hexadecimal machine code using the on-board calculator keypad. According to a MOS employee, the KIM-1 was much more well known among hobbyists than the Apple I, selling approximately six thousand units in a single year compared to the Apple I's estimated lifetime sales of less than two hundred units.[66] When Commodore Business Machines purchased MOS Technology in 1976, Peddle began designing a more complete computer around the 6502 that could be marketed toward novice users. In January 1977, Peddle unveiled the PET to the electronics industry at the Consumer Electronics Show.

Even if we set its black-box design aside, we can see evidence of Commodore's desire to dissociate the PET from hobbyist computing in its lack of engagement with American computing magazines and aggressive attitudes toward computer shops. This dissociation was likely driven in no small part by Tramiel, who stated publicly that he preferred to devote most of the Commodore's advertising efforts to the European market because he felt that the American market was "still geared to hobbyists" and therefore could not guarantee the volume of sales he desired.[67] Unlike Commodore's later appliance computer, the VIC-20, the PET was almost never advertised in American computing magazines. For example, Commodore had purchased only three advertisements in *BYTE* before it first advertised the VIC-20 in the January 1982 issue: one for the KIM-1 in November 1978, one for a PET software catalog in June 1981, and finally one for an upgraded, 4001 model of the PET in December 1981. Later reporting also suggested that Commodore had difficulties during the late 1970s securing relationships with American computer shops because the company would only sell through retailers with stores that fulfilled specific requirements, including filling out a formal application and paying a $2,500 deposit. Even those retailers with whom Commodore agreed to partner had trouble getting enough machines from the company to fill orders and were regularly impacted by Commodore's continual price adjustments.[68]

Public remarks from Commodore's employees similarly tried to dissociate the PET from hobby computing by promoting a contrasting rhetoric of usability stressing that computers should be immediately usable and useful to novices. When discussing the PET's design, Peddle repeatedly emphasized that those who would find it the most useful were people who knew very little about computing. In a series of interviews promoting the PET, he explained that the computer should be viewed as a new kind of appliance. The PET team "tried to make a product that is merchandisable by a normal retailer to the ultimate consumer. . . . For a sale to be made by an inexperienced retail clerk to an inexperienced customer the unit has to have immediate perceived value. The only way to have immediate perceived value is for the unit to do something the customer wants as soon as its plugged in."[69] Peddle wanted consumers to compare the PET to other appliances, not to other computers. When asked why someone who does not know anything about computers would decide to buy a PET, Peddle, according to one article, "expressed a sureness that when housewives see how simplified their lives, and challenged their minds can become with the addition of a computer, how can they resist?"[70] Like Tandy's executives, Peddle suggested that the PET would be seen as valuable in business contexts but likely only after users had discovered its utility at home:

> Q: Do you see that analogy as going further, do you see this becoming an appliance that is going to be widely used?
>
> A: That is our goal. We're doing everything we can to make it happen. In other words, the product's technical and marketing direction is to be a consumer item, with a secondary strong market emphasis on small business applications. But in addition to that, it is our intention to find ways to make this product useful to insurance salesmen, doctors, real estate salespeople—the classic professionals, people who have money, but more than that, people who are considered "thought leaders" in their community.[71]

Securing such widely recognized utility, he continued, would depend on having software applications available that would be seen as meeting real needs, adding that Commodore was planning to develop its own catalog of licensed software. Any developers who wanted their software included would first need to demonstrate that their software had been installed and put into use in a "financially rewarding environment."[72]

Peddle was also at times openly critical of hobbyists' construction of personalness as freedom, arguing that the idea of establishing a highly individualized relationship to a machine had limited appeal. Peddle stated explicitly, for instance, that "Commodore is not concerned with trying to replace 20,000 or so existing home computers" built by hobbyists: "If we didn't replace any of these, we'd hardly notice it in terms of the overall numbers."[73] Computers with open designs demanded too much time and attention from users for the modest benefits they offered. As a result, the average person still "has no concept that personal computing is here." A more universal approach to usability modeled after home appliances, he argued, is necessary if computers really are to be something found in every home and office: "People have been taught that computers are difficult to operate, that computers are things to be afraid of, not things to get warm and cuddly with. Therefore, what we've tried to do is to package the unit in such a way that it's as close to warm and friendly as we can get it; but it has value as a thing that does something."[74] According to Peddle, even the modest hardware setup required of the Apple II and TRS-80—connecting the keyboard unit to a display and ensuring the cassette recorder was set to the right speed and volume—was too much to ask of the average person. Appliances were things you just plugged in. The PET's all-in-one design would provide users with that experience. Moreover, appliances were not things you tinkered with. You took them to professional service departments if they needed repairs. The PET's design was thus intended to discourage hobbyist approaches: "We have really cut down your ability to screw with it. I hate pushing my competition, but if you really feel like you have to get in and mess around with your computer, buy an Apple."[75] In other words, Peddle believed that most people did not want a machine that demanded things of them, nor did they have any interest in computing beyond the tasks Commodore planned to support. This emphasis on computers being immediately useful, Peddle insisted, was a friendlier form of design than ones that insisted on personal control over every technical detail of their construction.

Compared to those written for the TRS-80, reviews for the PET in computer magazines were more critical of its closed design and at times openly concerned that it would pose problems for novice users. A May 1978 review by engineer Ralph Wells in *Kilobaud Microcomputing*, for example, concluded that Commodore's reluctance to provide technical information about the PET to users would undermine the appliance-like experience it envisioned. Although he praised the simplicity of its setup, he noted that it was impossible to diagnose problems when the machine appeared to operate incorrectly: "The big hang-up with bugs in my PET is that there is no service information provided. . . . [A]nd there are no complete spec sheets available for them."[76] Skeptical that a television or radio service shop would be able to fix such a complicated machine, Wells called Commodore's support line and was told that it would take two months for the company to service his PET. Deciding to try to fix it himself, Wells reported that he was ultimately successful but that the repair had required that he leverage his "$10,000 worth of test equipment and four years' experience with microprocessors." Wells further speculated that Commodore's appliance design may have less to do with making the machine seem "friendly" and more with keeping its circuitry "secret from competitors."[77] In a similar review that contrasts sharply with his discussion of the TRS-80, Fylstra questioned Peddle's claims that the Commodore could easily be serviced by local appliance electricians: "it is not yet clear, however, how or whether TV repairmen might be licensed to repair PETs under warranty. Presumably[,] experienced computer hobbyists could read the service manual and diagnose problems with their own PETs"; however, "how Commodore might react to this possibility is an open question," as those manuals were not readily available.[78] Nonetheless, both Wells and Fylstra responded favorably to its rhetoric of usability. The idea of a computer as easy to use as a television set or refrigerator, they seemed to suggest, would likely go a long way to dispel some of the negative perceptions in American culture about computing.

Reviews by novice users similarly found the concept of an appliance computer intriguing, but they often concluded that the model of usability the PET enacted was not as simple as Commodore's rhetoric of usability promised. Writing in *Money*, Peter Martin noted that he had frequently described computers as "the next major home appliance" in several articles but prior to this review had not yet used one himself. Characterizing himself as "every inch the ignorant layman," Martin suggested that he is exactly the sort of consumer that Commodore was envisioning in its promotions. Martin, however, found the PET "disconcerting," explaining that after "three hours of eyestrain I was able to accomplish what normally would take me five minutes with a $10 calculator." The machine came only with a small pamphlet for a manual, which he found to be incomprehensible and incomplete. Despite Peddle's claims that it could be plugged in and turned on easily, Martin could not figure out how to turn it on properly without calling Commodore's help line. Nor did he know what to do with it once he had gotten it to turn on. Ultimately, he settled on copying programs out of a BASIC handbook and some magazines that he had gotten from his son, but he was not impressed by the results.[79] In a similar piece published in *Newsweek*, Elizabeth Peer described her experience using a variety of appliance computers, including two variants of the Apple II and a Commodore PET. Despite having a background in math and science, Peer explained that after "a total of 70 hours with the computers" she "remained wholly incapable of utilizing them to suit the needs" she would consider buying them for.[80] Like the hobbyists, novice users writing about the PET praised Commodore's goal of trying to make computers easier to use but often concluded that the friendly, appliance-like experience promised by them was not realized in practice.

Although Commodore ultimately focused on promoting the original PET more in Canada and Europe than in the United States, both computing magazines and mainstream publications alike responded more favorably to the rhetoric of usability surrounding the PET than they did to the experience of using the machine. Tandy may have provided a retail environment for its machine that was similar to the one in which consumers bought appliances and had them serviced, but Commodore worked harder than Tandy to reframe home computers as more appropriate for the average person than for obsessive hackers. Tramiel himself would even state throughout the 1980s that the company's goal was to promote the idea of "computers for the masses, not for the classes." While Tramiel's comments can be interpreted as an attempt to position Commodore as a company in touch with the hobbyists' goal of upending technocratic control of computing, its advertisements remained more focused on promoting specific uses for computers. Beginning in 1980, Commodore expanded its advertising in the United States while maintaining its rhetorical commitment to the idea of immediate utility and a comfortable model of use. The next model of PET was presented in advertisements as "the great American solution machine," and Commodore's follow-up to the PET, the VIC-20, as "the friendly computer."[81] Although the PET was now a business machine, the idea that computers could be specially designed to play specific roles supported Commodore's earlier rhetoric of usability characterizing the PET as a machine that traded away the ability to tinker in order to make computing friendlier.

Atari 400 and 800

By the early 1980s, a second wave of appliance computers had entered the American market. Among them were the Atari 400 and 800. Atari's rhetoric of usability had more in common with Apple's than it did with Tandy's and Commodore's, however, in that it drew on countercultural tropes to position its products as a convenient way to participate in a social revolution through computing. Atari's catalogs and official magazines, for instance, are full of photographs that focus more on its users than on its products. The vast majority of computer advertisements during this time tended to portray white men fantasizing over the power this new technology would grant them.[82] By contrast, Atari represented its users as having diverse interests and identities: its users were not just businessmen and secretaries in corporate environments but college students, parents, and creative professionals. Their catalogs featured almost as many women as men and included far more images of nonwhite users. While Atari's computers are more often remembered via the aggressive antipiracy campaigns that the company launched in support of them during the early 1980s, it is important to recognize, too, the many comments made about Atari's reluctance to share publicly technical information about its system-specific features. In short, Atari was broadly concerned about controlling the types of software that users would have access to and took a variety steps to assert that control in order to maintain a public image of its computers as machines that made computing practical for and accessible by everyone. But despite these efforts, many users assumed it was simply an advanced gaming console and ignored any efforts to portray these machines otherwise. In this respect, the conflicting rhetorics of usability that circulated around the Atari 400 and 800 thus serve as a reminder that the social construction of personal computing can lead users and designers to adopt dramatically different views of technology. Rhetorics and models of usability often serve to normalize user behavior, but a designer's power is not absolute.

Unlike Tandy and Commodore, Atari benefited from the fact that the American public was already somewhat familiar with the idea of appliance computing as it began its sale of its personal computers in early 1979. An early review of the Atari 400 and 800 by John Victor, for example, declared that these machines represented a "third generation of microcomputer" that took inspiration from Apple, Tandy, and Commodore while also introducing some important hardware innovations to provide consumers with a "true personal and home computer system." Victor began his review with a brief history of personal computing that explained how the first generation of machines modeled after the Altair 8800 "were not designed with any particular purpose." Whereas advocates of kit-based computers claimed that the potential of their machines was limitless, Victor characterized their general purpose design as a response to market uncertainty: "They had to be designed for all possible configurations with plenty of slots for memory boards, large power supplies, cooling fans, etc. First generation computers were expensive, and many users were paying for features they did not need." Their emphasis on personalization, Victor implied, had made it difficult for consumers to conceive of clear uses for them. Apple, Tandy, and Commodore had taken important steps to make the second generation of personal computers—the Apple II, TRS-80, and PET—more affordable and to present a clear sense of utility to the public by narrowing and focusing their designs around more specific purposes. Atari's machines took this process a step further, producing a system that was "excellently suited for the educational and recreational interests of the consumer market."[83] Another review by Chris Morgan similarly suggested that Atari had innovated the idea of appliance computing by creating a "hybrid computer" that "looks and acts like a video game, but which also has the features of a personal computer." Gaming consoles had already shown that computing devices could be designed to be immediately and universally usable by novices, so it seemed logical to design a home computer that borrowed on the design features of gaming consoles if user-friendliness were a top priority. Morgan expected that other manufacturers would soon copy Atari's model of usability.[84]

Prior to 1979, Atari had been known primarily as a gaming company, and many early reviews considered the 400 and 800 in this context, treating them as gaming consoles that incorporated elements of personal computers. In his review for *Creative Computing*, for example, Ted Nelson did not carefully describe and comment on each component of the computer, as was the convention in most computing magazines. Instead, he recounted a demonstration he saw of the 800 and then proceeded to rave for several paragraphs about how the graphics made him feel as if he were being taken into outer space.[85] Another review, written by Frank J. Derfler Jr. for *Kilobaud Microcomputing*, took the form of a letter detailing the author's experience binge drinking and playing video games with his friend. Derfler concluded that it was fun and well-designed but unfortunately not well-designed enough to continue working after his friend spilled moonshine all over it near the end of the evening.[86] More serious reviews emphasized that it was far more than "the ultimate in home video games" because "each unit is, in fact, designed as the heart of a computer *system*."[87] In other words, the Atari 400's and 800's best feature was that they could be used not only to play games but also to write them.

While Atari did at times acknowledge users' interests in gaming, Atari's executives also frequently drew on rhetorics of universal and immediate usability that tried to frame appliance computers as machines that anyone could use for more serious purposes. In an interview, Atari's vice president for sales and marketing for personal computers, Conrad Jutson, explained that the goal of the Atari 400's and 800's design was "to take away whatever apprehensions a first time user might have and help him or her feel good about interfacing with our product." Users, he continued, did not care about technical specifics. In fact, basing sales pitches on a description of technical merits might "scare the consumer off by making it so he or she has to have a double E [electrical engineering degree] or be a computer programmer to utilize the full capabilities of a personal computer. . . . With Atari computers, you don't have to stop and think before you use them."[88] Jutson emphasized that more people would be more willing to purchase a computer if they were able to visualize in more concrete terms how it would fit into their lives. Rather than ask does it have the right kind of processor or disk drive, they are more likely to ask "what will it do for me?"[89] Although Atari's initial advertisements in 1979 and 1980 were full of copy inviting consumers to evaluate the 800 on its technical merits, the company shifted after 1981 toward more minimalist advertisements that appeared to take Jutson's advice. Starting in May of that year, it began running advertisements that simply referred to "computers for people" in large type that took up half the page (see figure 3.1). Some advertisements in the campaign included endorsements from successful creative, business, and scientific professionals touting the productive potential of Atari's computers. Atari relied on these advertisements in conjunction with its catalog photos showing a racially diverse, mixed-gender, multigenerational group of users as a way to persuade consumers that the 400 and 800 were meant for more than gaming. These computers had finally opened up the personal computer revolution for everyone. These were machines that were built for real, ordinary people and that would help them achieve whatever goals they may have had. Atari's

Figure 3.1. The user-friendly "Computers for People" marketing slogan used in advertisements for the Atari 800. This image comes from its first appearance in the May 1981 issue of *BYTE* magazine.

computers could be put to use easily and immediately, and they would empower users in a variety of real contexts.[90]

Like Commodore, Atari also hoped that their appliance computers would first see use in the home before being adopted by business users as they became more popular. Reporters that spoke directly with Atari users, however, discovered that many had trouble seeing Atari's computers being used in office settings. Discussing conversations he had had with Atari executives in an interview with *InfoWorld*, the head of the San Francisco Bay Atari Users Group, Clyde Spencer, suggested that Atari viewed gaming as a friendly way to help novices become comfortable with using a computer. In the long term, Atari expected to grow more around business software: "Management . . . really want[s] to hook the consumer on games. Then, once the machine's in the house, if they get tired of games, they'll do other things with it."[91] But Spencer added that based on what he had observed within the users group, the "other things" that Atari users tended to do with their 400 and 800 computers mostly involved programming their own games. In addition to portraying computing as a fun and inclusive activity, Atari's marketing materials also detailed business uses and specifically touted the availably of VisiCalc for the 800, but other users who were interviewed seemed surprised when asked about the company's views on productivity software. They commented further that it was not a sense of "usefulness" that led them to buy an Atari and generally agreed that "business applications were Atari's weak software point." Many simply assumed that Atari had "consciously made the decision not to compete with IBM, Xerox, HP, and Apple for the small-business market."[92] By the 1980s, the experiences of Atari's users were contributing to the increasing perception that appliance computers, when compared to more expensive options, were meant primarily for less serious activities like gaming.

Atari's attempt to portray the 400 and 800 as computers that could be used for serious purposes was further hindered by the fact that both machines utilized cartridge media resembling the kind used for the Atari 2600 game console. Most personal computers at that time stored their essential software on read-only memory, or ROM, chips. ROM chips for BASIC interpreters, for example, were soldered to the main board of most personal computers. For open architecture personal computers, like the Apple II or IBM PC, new ROM chips could be installed via expansion boards that slotted into the main board's bus as semi-permanent upgrades. Unlike software stored on cassettes or floppy disks, software stored on ROM chips would be accessible as soon as the system was turned on. Jerry Lawson developed the first commercially viable modular ROM cartridge in 1976 for use with the Fairchild Channel F game console; however, modular ROM cartridges could also be found in more professional contexts such those Texas Instruments developed for use with its programmable calculators in 1977.[93] Atari had already begun manufacturing them in 1977 for the 2600.

While ROM cartridges were convenient, they also offered Atari a significant advantage in the software market. Floppy disk storage had made it easy to make and share copies of software; however, ROM cartridges made doing so very difficult for most users.[94] On the other hand, they also introduced a barrier to software developers who wanted to publish applications for Atari's systems. Independent, third-party development for the Atari 2600 had not been possible until 1979, when ex-Atari employees reverse engineered the 2600's cartridge system and founded their own publishing company.[95] By the early 1980s, Atari had developed a reputation as a company with an aggressive attitude toward third-party software developers. As several journalists observed, Atari's frequent legal claims regarding piracy were often dubious, targeting primarily smaller software developers that the company felt were developing software that too closely resembled their own products.[96] Atari's control over the software published for its computers was not perfect, however. Even though many smaller third-party developers were unable to manufacture their own cartridges, they could still publish software on tapes or disks. Nonetheless, cartridge-based media did help Atari exert a fair amount of control over the software ecology that took shape around its computers.

In addition to developing a cartridge system, Atari also withheld technical information from the public in order to press software developers to sign publishing agreements in exchange for access. When reviewers were considering the 400 and 800 as computers for novices, they described the included handbook as "explicitly detailed," praising it as a tool for beginners looking to move beyond BASIC into "sophisticated systems programming."[97] When they considered them as machines for more experienced programmers, however, some noted that Atari's manuals excluded instructions on how to program for its propriety graphics processing chips. For example, when Nelson reflected on the software demonstration for the 800 that he had witnessed in light of his understanding of how other systems rendered graphics, he came to the realization that the machine's graphics chipset was so unique that he would not be able to create similar software without access to a detailed overview of its hardware. In his review, he claimed that Atari "won't tell you how" to the create the same effects in your own software: "Everything is under wraps. Oh, of course you can program the 6502 chip, that's in there, the same stuff as Apple. But that *other stuff*, those mysterious peek and poke locations . . . are a deep dark secret." Nelson wondered, much as Wells did with the PET, if Atari's lack of documentation was a ploy "to hobble potential software rivals. . . . Atari is being co-operative [only] with independent software vendors, *provided* they don't tell [anyone else] how it works."[98] More subdued reviews wondered if maybe the gaps in Atari's documentation were just an oversight, suggesting that that the reference manuals had to "have been prepared early in the machine's history" because they were "somewhat sketchy" and did not seem to describe some of the more advanced programming features well.[99]

Although many users spoke highly of the 400 and 800, Atari was not able to realize a long-term acceptance of its computers as usable and useful for anything more than gaming. Michael Tomczyk has speculated that Atari's attempt to control its platform may have ultimately prevented the 800 from being seen as a more serious machine in spite of its console design. In attempting to "blackmail" software developers by withholding information, the company "alienated the fiercely independent hobbyist / programmer community, and as a result many serious programmers started writing software for other machines instead. . . . The only programmers who remained loyal were game programmers."[100] Atari's inability to convince users of the potential business uses for the 800 also shows how appliance computing was coming to be seen as a less serious approach to personal computing. While designers can exercise significant rhetorical influence over our understandings of the models of usability they produce, that influence is constrained by the larger discourse they are received through.

Texas Instruments TI-99/4A

When mentioned at all in prominent histories of American personal computing, Texas Instrument's appliance computers are typically discussed only as part of the "home computer shakeout" of 1983 that effectively marked the end of early appliance computing.[101] Yet it is estimated that Texas Instruments was selling approximately thirty thousand TI-99/4As per week by the end of 1982. Before Texas Instruments ceased its manufacture in November 1983, many journalists believed that it had the largest install base of any personal computer in the United States. The handful of extended looks at Texas Instruments' participation in personal computing published during the 1980s are thus written with the stated goal of understanding the mistakes made by its executives that caused it to vanish so suddenly from the market at the height of its popularity. As I discuss in this section, the rhetorical history of Texas Instruments' computers is surprising not because of the dramatic financial losses the company incurred but because it associates appliance computing with a vague promise of friendliness without clearly outlining what specific utility it might have. Unlike the other companies considered in this chapter, Texas Instruments effected a sense of user-friendliness not by giving people reasons to buy a computer but by trying to eliminate reasons for them not to. In doing so, it developed a rhetoric and model of usability that did not try to mask its aggressive control over software like other manufacturers but instead pushed users to avoid considering the consequences of that control.

The TI-99/4A was initially proposed as a cheaper revision of Texas Instruments' first computer, the TI-99/4. The TI-99/4 was released in 1979 and was celebrated for being the first home computer built around a 16-bit processor. Some journalists quietly hoped it would push other manufacturers to move on from the 8-bit processors that had defined personal computing since the Altair 8800.[102] Despite its technical advantages, the

TI-99/4 sold poorly. Compared to other commercially available personal computers, there was almost no software available for the TI-99/4 because few programmers were willing to take the time to port their applications to its 16-bit architecture. After disappointing sales, Texas Instruments would replace the TI-99/4 with a lower-cost machine, the TI-99/4A, which debuted in 1981. In 1982, Texas Instruments claimed to have sold five hundred thousand TI-99/4A computers and stated that it expected to sell anywhere from one to three million the following year.[103] Part of its strong sales owed simply to its low price. The TI-99/4A debuted at $525 and was eventually sold for as low as $99 after rebates.[104] Executives at Texas Instruments planned to make up any lost revenue from the price drops through sales of software and accessories. Despite the fact that the TI-99/4A quickly became a top-selling computer system in the United States, Texas Instruments was unable to realize the sales volume it needed to make the TI-99/4A profitable. In the beginning of 1983, Texas Instruments' computer division reported $100 million in losses and in November announced that it would discontinue the TI-99/4A, at which point the company withdrew almost entirely from the American personal computer market. Although finances played a role in this decampment, the company's seemingly strange moves can be better understood through an examination of how it positioned itself against other manufacturers of appliance computers.

Compared to the rhetorics of usability developed by the other companies I have already discussed, Texas Instruments' is somewhat paradoxical. The company's advertisements did not address the topic of how or for what purposes the TI-99/4A should be used, and executives largely avoided engaging with the technology press. The company did not just generally ignore American computing magazines, as Commodore did. Instead, it deliberately chose to exclude its appliance computers from the advertisements that it published in them throughout the early 1980s. Beginning in 1980, Texas Instruments purchased full-page advertisements for its calculators in almost every issue of *BYTE*.[105] During this time, the company also purchased a number of smaller, half-page advertisements announcing that it was hiring software engineers.[106] It also purchased full-page advertisements for its first-party software publishing program, inviting readers who were interested in selling software they had written for the TI-99/4A to apply for a licensing agreement.[107] But it did not purchase a single advertisement in *BYTE* for the TI-99/4A itself until September 1982, which featured a photograph of Bill Cosby standing next to the keyboard unit, leaning against a monitor that rests atop the machine's PEB, beneath the words "This is it. This is the one." The photograph takes up three-quarters of the page, with the remainder filled by copy that lists its features but does not point to potential uses apart from noting that it was intended for the home, not for businesses.[108] Texas Instruments ran no variations of the advertisement in *BYTE* and ceased advertising the TI-99/4A in the magazine after December 1982. This purchasing strategy suggests that Texas Instruments had little interest in attracting buyers from technical backgrounds. They were more likely to write software for the TI-99/4A than they were to use one of the machines themselves. Texas Instruments would later advertise some of the software packages it published for the TI-99/4A in other magazines—notably *99'er*, an American publication devoted to both of Texas Instrument's appliance computers—and those advertisements would focus on the specific uses of those applications. But as to the machine itself, American advertisements for the TI-99/4A in computing magazines offered almost nothing apart from a large photograph of it.

Texas Instrument's dismissal of expert users and its vague representation of its computer in advertisements were part of a broader strategy to make consumers comfortable buying the TI-99/4A as an impulse purchase. Like Tandy, Texas Instruments did not want its computers sold in specialized shops. However, even a general consumer electronics store like Radio Shack was too specialized for Texas Instruments. It wanted to transform personal computers from something that customers needed to research and ask questions about into something they could buy without much thought other than whether the price was right. The company's decision to ignore computing magazines was part of a larger strategy to prioritize "the kind of stores that already carried the company's pocket calculator, stores like J. C. Penney and Sears and Montgomery Ward."[109] Tandy, at least, took steps to train staff in computer sales and service and by the 1980s would even offer classes to users through its more specialized "computer center" outlets. Executives at Texas Instruments, on the other hand, believed that their consumers would not be bothered by the fact that they would be purchasing from stores that had little training or experience with computers: "You couldn't sell a home computer in a computer store. Computer stores were meant for people who already knew something about them or were serious enough about them to spend several thousand dollars on one."[110] In fact,

Texas Instruments seemed to have had little interest in explaining to consumers what they would use a computer for or why they should buy one. It planned to rely almost entirely on the general air of excitement in the market:

> The TI machine, on the other hand, was going to be the first computer designed for Everyman. Did Everyman need—or even want—a computer in this home? . . . TI would put out a computer just powerful enough to entice the average person to take the plunge—no word processing, but plenty of education programs for the kids—yet inexpensive enough that the plunge wouldn't break the bank. On the basis of price alone, TI thought, the machine would sell. Convincing people that they needed it could come later.[111]

If the "hard part about selling a home computer is that . . . it has no immediately recognizable purpose," then Texas Instruments moved to sidestep that problem entirely.[112] Instead of explaining its usefulness or utility, Texas Instruments' advertisements just showed a wanly smiling Cosby who suggested that a decision had already been made for them: "This is the one."

While Texas Instruments ignored computing magazines, a handful of them did print reviews of the TI-99/4 and TI-99/4A. Typically, reviewers described both machines as technically superior to other appliance computers available to consumers, even better than the base models of general purpose computers like the Apple II or IBM PC. Yet they also expressed concern that both machines enacted models of usability that were closed in ways that may at a glance have seemed similar to, but in practice far exceeded, those of other appliance computers.[113] While Tandy, Commodore, and Atari had to varying degrees been reluctant to support third-party hardware and software developers, key components of their computers were either manufactured from third parties or available for sale separately. For example, Commodore owned MOS Technology, but the MOS 6502 microprocessor was still itself sold as a separate, well-documented product. Thus, while Commodore and Atari may not have initially released comprehensive technical manuals for the PET and 800 that included complete hardware instruction codes, software developers could use documentation for the 6502 as a starting point. The same was true for the TRS-80, which used a microprocessor manufactured by Zilog. The TI-99/4 and TI-99/4A, on the other hand, used the TMS9900, a processor manufactured by Texas Instruments and found only in its own machines. Whereas anyone could write software for the TI-99/4 and TI-99/4A in BASIC, writing software that took advantage of the unique features offered by its processor and proprietary graphics chipsets was only possible if one obtained documentation directly from Texas Instruments, which required signing a licensing and royalty agreement.[114]

In addition to having less publicly accessible hardware information, Texas Instruments also put pressure on third-party software developers both technologically and in the courts. Like Atari's computers, the TI-99/4 and TI-99/4A also had a cartridge-based ROM software system built into its keyboard unit. While some TI-99/4A software was sold on cassette tapes and floppy disks, Texas Instruments openly pushed third-party developers to create software on ROM cartridges "to minimize opportunities for software piracy." However, Texas Instruments required third-party developers who signed up to create software for the cartridges to purchase a minicomputer development system, "a hardware investment . . . on the order of $50,000 to $100,000."[115] By 1983, third-party developers had found ways to manufacture software cartridges for the machine without relying on Texas Instruments' expensive kit. The company responded by announcing that it would revise the TI-99/4A's hardware in a way that would enable it to check for an authorization chip when accessing a software cartridge, ensuring that all future TI-99/4As would only be capable of running cartridges manufactured by Texas Instruments.[116] In the meantime, Texas Instruments also threatened legal action for "patent infringement" against those third-party developers who did not agree to license their software through the company and surrender 90 percent of their sales revenue. As a result, many games and applications with names that might have been recognizable to first-time buyers and that were commonly available on other systems were not available for the TI-99/4A.[117]

Texas Instruments leveraged the rhetorics of appliance computing to put a vaguely friendly face onto an aggressive and predatory model of usability. Reporters interviewing new buyers observed that many who encountered the TI-99/4A and saw its "tempting price" felt that their purchase was "virtually inevitable." As the price for the TI-99/4A began to drop lower and lower, the company had to make up the losses it took on each unit. While its aggressive tactics with respect to software developers represented one approach to recover losses from hardware sales, another was to design store displays to encourage new buyers to purchase add-ons that far exceeded the cost of the system itself. Customers who were impressed by a demonstration model of the TI-99/4A that they saw on display at Sears—which retailed at $525 and was often cheaper with rebates—would eventually discover that they would need to spend about $530 more on add-ons to recreate in their homes what they had experienced with the demonstration model.[118] Market research studies indicated that the impulse purchase strategy that Texas Instruments relied on often did not translate into a long-term interest in computing among novices and likely contributed to the negative portrayal of appliance computers in early histories of personal computing. As one person interviewed explained, "I got my father-in-law a Texas Instruments 99/4A, which turned out to be good for about two months. He couldn't find things to do with it. He couldn't understand what memory was, so he typed in an entire program every time he wanted to use it instead of storing it on [cassette] tape." The study found that 39 percent of people who had bought a computer had stopped using it within six months. The majority of those who had stopped using their machines were people who had purchased cheaper appliance computers.[119]

Conclusion

The commercial failure of early appliance computing would ironically pave the way for the long-term acceptance of its associated rhetorics and models of usability. Although Texas Instruments had reportedly always wanted to keep its entry-price low and make up revenue through add-ons, by early 1983 it was engaged in a price war with Commodore and dropped its prices even lower. The conflict between these two companies eventually forced Tandy and Atari to lower their prices as well. The "home computer shakeout" that followed resulted in Atari drawing back and Texas Instruments beginning to exit entirely from personal computing. Many reporters conjectured that the market for appliance computing might collapse entirely by the end of the year.[120] Although some reporters feared that the price war might amplify existing concerns about the usability and usefulness of personal computers in American culture, more optimistic observers suggested that consumers were turning their attention away from machines they saw as useful only for "game playing [and] toward more powerful computers with greater capabilities." The shakeout "might actually restore in consumers' minds the computer's perceived value that had been damaged in the fierce price cutting."[121] Even though the cheaper appliance computers resembling those manufactured during the late 1970s and early 1980s became associated primarily with video games after the shakeout, the rhetorics of usability that circulated around them promising immediate utility would remain influential within the American personal computer industry. As I discuss in the next chapter, Apple would adopt many of the ideas associated with the early appliance computers for its Macintosh, presenting its black-box design and transparent graphic user interface as a solution to concerns about a looming computer literacy crisis.

Another important consequence of early models of appliance computing was the growing association between user-friendliness and proprietary control. As I discussed in the previous chapter, hobbyist users by 1977 had already begun to revise their narrative of the personal computer revolution in ways that incorporated the rise of privately and centrally managed standards. Companies like Tandy, Commodore, Atari, and Texas Instruments helped to similarly persuade novices that their control of software ecosystems was necessary to realize models of immediate and universal usability. While manufacturers of appliance computers did not explicitly suggest there was a tension between computing and culture, they nonetheless implied that the personalized computing hobbyists emphasized was not a mainstream interest and that focusing on it distracted from the more pressing concern that most consumers often could not understand what made computers useful to them. The loss of agency that appliance models of usability entailed was thus presented to users as a selling point. You would not need to open the computer yourself because service staff at Radio Shack would handle all of the more technical aspects of your machine for you. There was no need to worry about software compatibility because the Atari-branded cartridges were guaranteed to work with your system. Manufacturers of early appliance computers assured consumers in various ways that they would manage the more complicated aspects of computing on users' behalf. All consumers had to concern themselves with now was the question of what computing would do for them.

The important lesson of this chapter is that user-friendliness often represents far more than simply a commitment to making computers easier to use. The same features that a rhetoric of appliance usability makes attractive to consumers are often ones that offer manufacturers significant market advantages. While there is a broadly shared sense of what counts as "user friendly" in American computing culture, this chapter reminds us that the specific ways that software designers realize user-friendliness matter. The developers of the Mac operating system and those of the Windows operating system do not just disagree on aesthetics or whether the feel of the dock is more natural than that of the start menu. They disagree on how we should think about and behave when using their technologies. We need to begin looking at contemporary digital culture more closely with this lesson in mind. Today's walled gardens, forced updates, cloud services, and copy-protected media systems are pitched to us as necessary to maintain the ease and convenience we have been told we should expect out of personal computing technologies. As Emerson argues, models of appliance usability have transformed personal computing in a way that deprivileges creative practices in favor of consumptive ones.[122] But it is important to recognize, too, that today's appliance models of usability afford their designers broad control over the media ecosystems their platforms support. They can decide to ban media or software applications from their platforms entirely or force third parties to change their software to conform to their own aesthetic, ethical, or political standards.[123] This power may not be necessarily viewed as harmful by the general public, especially when it is used for benevolent ends like banning media developed by hate groups or that promotes violent behavior.[124] But often, platform policies are not enforced consistently. Platform developers make exceptions to their own policies for a variety of reasons and usually only act to address the problems those exceptions cause after an overwhelming amount of media attention makes Big Tech's corporate lawyers or stockholders nervous. Moreover, as I have shown in this chapter, hardware and software developers have historically done a good job positioning their technologies in ways that can deflect or at least delay public outcry over the consequences of their designs. By and large, the rhetorics of appliance computing have led us to accept the benevolence of Big Tech's stewardship of digital culture because we now expect and depend on the ease and convenience of their models of immediate and universal usability even as we brush up against the constraints it places on our agency.

Of course, there are a few steps between the appliance computers of the late 1970s and early 1980s and today's ultra-transparent walled gardens. As I discuss in the next chapter, the growing excitement over appliance computing in the United States paralleled an increasing frustration over the growing expectation of computer literacy. As David D. Thornburg and Betty J. Burr argued in a 1980 column in *Compute!*, what computers of the late 1970s had shown was that the personal computer revolution needed to be reframed into "friendly revolution." Computers needed to be "designed to respond to the needs and desires of people on people terms." Machines like Tandy TRS-80, Commodore PET, Atari 800, or Texas Instruments TI-99/4A may not have gotten everything right, but they did show that "as revolutionary as the enabling technology has been, the mere existence of the personal computer is insufficient to give everyone access to all the things computers can be used for. In order for this technology to move into people's homes, the interface between computers and people has to be improved to the point where the average person can operate the computer as easily as he or she can operate a color television." None of the appliance computers were truly as easy to use as an appliance, but the rhetoric around them had introduced interesting ideas for future design. Personal computing needed a new type of interaction, one that would "humanize" the computer and make it "sufficiently natural to users" so that it "becomes a transparent facilitator between the user and the goal of interaction—be that playing a game, watering the lawn, looking at a stock portfolio, etc."[125] The idea of appliance computing—of machines whose utility was attendant on a narrowing of purpose via a black-box design—appeared to some to have been more successful than the specific machines associated with it. The failure of appliance computing to realize its promise of user-friendliness, in short, helped to provide a sense of exigence for new rhetorics of usability focused on transparent design.

Chapter 4

IBM, Apple, and a Computer Literacy Crisis

In many representations of early personal computing culture in the United States, Apple's *1984*-inspired commercial looms large. Nationally broadcast during Super Bowl XVIII, the commercial implied not only that the most important goals of the personal computer revolution had yet to be realized but also that Apple now remained its lone vanguard. Only Apple could put an end to IBM's efforts to force its corporate culture of conformity onto the rest of the world. Popular memory of the conflict between Apple and IBM that unfolded during the 1980s is often strongly sympathetic to Apple's framing, likely because of the continued privileging of hobbyist narratives discussed in chapter 2. Yet if we step back and look at the conflict between IBM and Apple anew, we can see that it was not about digital freedom versus computational conformity but about competing definitions of computer literacy. Across conversations on usability taking place in American computer magazines in the early 1980s, the consensus that transparent design was the only way to achieve user-friendliness begins to crystallize in debates over the skills and knowledges users needed to develop to see personal computers as usable and useful.

Both IBM and Apple developed rhetorics of usability that appeared to respond to growing concerns that the average American would never be able to navigate an increasingly computerized society. However, in tracing the development of these rhetorics, it is equally important to consider the different definitions of computer literacy they intended their models of usability to support. Even were we to accept Apple's framing of its conflict with IBM a , primarily about freedom versus conformity, viewing that conflict with each company's definition of computer literacy in mind shows that

Apple pushed users to conform to a radically narrowed understanding of computing, while IBM aimed to cultivate expertise among users and to support a wide range of computing practices. Whereas Apple suggested with the Macintosh that the company understood what users wanted better than users themselves, IBM at least initially claimed that it could not predict the future direction of personal computing and would leave it to a newly computer literate public to determine the path forward. In this chapter, I examine the rhetorics and models of usability associated with the IBM PC and the Apple Macintosh to show how user-friendliness came to be strongly associated with transparent design during the 1980s as both companies positioned their machines as responses to growing concern about a looming "computer literacy crisis."

Although journalists and other observers writing about the personal computing industry in the late 1970s and early 1980s frequently discussed the importance of computer literacy, they rarely took the time to define it. In synthesizing their writings, I show how many discussed computer literacy in terms similar to those Annette Vee uses in her description of programming literacy as a sociomaterial phenomenon defined by material skills and technologies as well as by the specific conceptual understandings and social norms of computer use within a given context.[1] In other words, these writers understood that what "counts" as computer literacy is a product of both the model of usability enacted by a specific device and the rhetoric of usability that informs our interpretation of it.

Vee's work documenting the history of programming education in the United States further shows how early discussions of computer literacy treated programming as a broadly practical skill that would soon be required across most professional contexts. For most of the first decade of personal computing, the idea of "computer literacy" was synonymous with "programming literacy." While Vee concludes that this association has only strengthened over time, I argue in this chapter that user-friendly approaches to usability have historically worked to maintain programming literacy as a separate, specialized domain distinct from a more general form of computer literacy. However, I do not believe that our arguments are mutually exclusive. Indeed, the design of most modern programming languages has been informed by the same information-hiding practices that influenced the principles of transparent design. These practices allow programmers to focus on developing software at a "high level," without requiring them to explicitly define algorithms or directly manage system states at a “low level.” In other words, many programming languages in use across a variety of professions today are structured similarly to conceal core computational processes so that coders can focus solely on the phenomena they are modeling or the data they are processing. Some of the arguments I make in this chapter could with some revision could be applied to the politics of code reuse, programming pedagogy, and the management of collaborative programming projects.

The distinction between an abstract understanding of idealized theories of technological function and a practical understanding of how those principles are realized or represented in specific technologies is crucial to making sense of debates over computer literacy in the early 1980s. This distinction is similar to what Andrea diSessa describes as the difference between “computational literacy” and “computer literacy.” Whereas the former provides a foundation for understanding how computational media restructure communication—and thus could serve as the foundation for new approach to education—the latter more narrowly focuses on the operation of computers. Consequently, diSessa’s computer literacy is also in practice a “consumer literacy” in the sense that it encourages us to conform to a set of predefined, intended behaviors when we are using digital media rather than seek to create our own.[2] While the journalists, executives, and other participants in conversations documented in the material I examined do not use these terms, I follow diSessa’s example in this chapter in order to highlight the similar distinction that they often drew between those skills and knowledges they saw as appropriate for users and those they saw as best left to expert technologists. Those who supported IBM’s definition of computer literacy worked to develop a rhetoric of usability that would make computation legible to novice users. However, others felt that widespread adoption of IBM’s model of usability would exacerbate an ongoing computer literacy crisis. Drawing on the counter rhetoric of usability that these contributors to early computing magazines began to develop, Apple positioned its Macintosh as able to support a form of computer literacy that allowed computers to be usable and useful without reference to computational concepts.

Ultimately, however, is it important to remember that computer literacy is more than a functional literacy. As Cynthia Selfe’s and Stuart Selber’s writings remind us, our functional skills with and conceptual understandings of digital media necessarily inform our ability to leverage their rhetorical affordances and to critically engage with the sociocultural contexts of computing.[3] In working to foster distinct definitions of computer literacy, IBM and Apple were in this respect also pushing users to adopt favorable interpretations of their models of usability, both of which were influenced by a variety of concerns beyond ease of use. Both companies wanted users to adopt a language, way of thinking, and basic skillset that made their models of usability feel like the future of computing. It is important to recognize, too, that the computer literacy crisis that observers described was like other literacy crises in the United Stated in the sense that it was about more than simply a skills gap. John Trimbur has observed that literacy crises in the United States have served as "ideological events": "strategic pretexts for educational and cultural change that renegotiate the terms of cultural hegemony, the relations between classes and groups, and the meaning and use of literacy."[4] Despite IBM's reputation for maintaining tight control over technology, the rhetoric of usability around the PC implicitly drew on the idea of digital freedom by suggesting that anyone was capable not only of understanding computation but also of recognizing its broader sociocultural implications. Apple, on the other hand, pointed to a tension between computation and culture. The rhetoric of usability it moved to associate with the Macintosh suggested that computational concerns were inherently alien to everyday experience and so were best managed by benevolent innovators who prioritized the needs of novices. If we recognize that rhetorics and models of usability necessarily shape our computer literacy, then we must also recognize that design is a process through which hardware and software developers seek to inculcate within users specific languages, skills, and understandings. And if the goal of design is a user-friendly technology, then those specific languages, skills, and understandings should make a given technology seem as if it fits intuitively into users' beliefs about how computers should work or the roles that they should play in our culture.

I begin this chapter by first examining discussions taking place in American computing magazines in the late 1970s and early 1980s about a looming computer literacy crisis. These discussions gave rise to the idea of a conflict between computation and culture, a conflict that could only be resolved, it was suggested, if a concerted effort were made to provide Americans with the right kind of computer literacy training. In 1981, as these conversations were unfolding, IBM introduced the PC. The PC's open design was one that many would copy, becoming a de facto standard for personal computing in the United States by the end of 1982. IBM would develop a rhetoric of usability for its documentation and open design that reframed the PC's hurried development process as one that carefully foregrounded the needs of novice users by offering a structured program for developing computer literacy. Many, however, saw the PC's model of usability as differing little from others already on the market and suggested the conflict between computation and culture could only be resolved if we developed new technologies that did not require a sophisticated computer literacy so that a basic competency could be realized more immediately. After discussing reactions to the PC, I then look at how Apple drew on rhetorics and models of usability from the late 1970s to position the Macintosh as a response to criticism of the PC. While many reviewers noted that the Macintosh's design radically narrowed the scope of computer literacy in ways that deviated from early 1980s norms of usability, Apple was able to reframe its design as resolving the tension between computation and culture by contrasting it with the PC.

"A Crisis of Epidemic Proportions"

By the 1980s, rising sales of lower-priced "appliance computers" had helped to change the assumption that personal computing was just a niche hobby. As discussed in the previous chapter, companies like Tandy, Commodore, Atari, and Texas Instruments had accompanied their products with rhetorics of usability that promised easy integration of computing into Americans' working and home lives. Descriptions of appliance computers in the 1980s, unlike those of personal computers in the mid-1970s, emphasized that they prioritized the experience of the novice user by providing them with immediate utility. Each company also introduced their own proprietary standards that afforded them a degree of control over the media ecologies that formed around their machines. While their rhetorics of usability had a lasting influence on what consumers would expect from personal computing, their models of usability also contributed to ongoing problems related to hardware and software incompatibility. In some respects, compatibility was less a concern for the kit-based machines that the hobbyists had favored because many computers built around the S-100 bus that the Altair 8800 had introduced also used the Control Program for Microcomputers (more commonly known as "CP/M") operating system, which meant that despite some significant differences across kit models or even individually customized machines, many hobbyists could share data and software among themselves.[5] Appliance computers, on the other hand, were purposefully designed to have limited or no compatibility with a competitor's products. Each had proprietary expansion standards, disk operating systems that were outwardly similar but implemented different file system structures, and even supported different "dialects" of the BASIC programming language. While appliance computing helped popularize the idea of easy-to-use computers, in practice it also introduced many new problems that complicated digital culture and frustrated users.

Among those problems was a broadly shared sense that using these new, "friendly" computers was not as easy as using the familiar appliances they were compared to by their manufacturers. Several journals and other early observers of the industry suggested that buyer's remorse among novice users was growing. David D. Thornburg, for example, claimed in a 1981 column that acquaintances regularly told him about how they bought a computer with the idea that it would save them time or money only to discover that they "have little understanding of the effort needed to make the computer do truly useful things." Computers were "being sold to many thousands of people who have no idea what they were getting into." Most people "could save a lot of grief by flushing the money down the toilet instead."[6]

Similarly, Robert Cowen suggested in 1981 that the computer companies and the popular press had been talking for several years as if a technological revolution had come to fruition. But in practice, very little about personal computers had changed apart from the fact that they were now easier to buy. The software available to consumers felt "tedious" and the tasks they assisted with were "trivial." Despite their growing popularity, Cowen suggested that personal computers were still "foreign and baffling to people outside the . . . subculture." Computer hardware manufacturers and software developers had embraced the language of appliance computing and user-friendliness as sales strategies but in practice did little to realize models of usability that lived up to their rhetoric. Actually achieving the kind of broad cultural and social benefits of personal computing that hobbyists imagined would require that manufacturers had a "deep understanding of how people relate to computers and master them," an understanding that would enable them to design their "services, software, equipment, and instruction manuals" appropriately.[7] Some months later in response to letters criticizing him for questioning the significance of the growing industry, Cowen clarified his point, noting that the problems facing personal computing were cultural and social, not technical: it took "widespread literacy for the printing press to make its impact, to say nothing of the growth of the publishing industry. The home computer awaits a similar rise in computer literacy and an information industry that caters to it. These will be the real agents of change."[8] Appliance computers had begun to put computing power in the hands of the average person, but there was still a broadly shared sense that, outside of work-related contexts, they were not usable or useful for anything more than tinkering in BASIC or playing video games.

Some observers of the rapid growth of personal computer industry even went so far as to assert that the United States may soon face a "computer literacy crisis." One of the earliest mentions of a computer literacy crisis can be found in a 1979 National Science Foundation–funded study of science education outcomes. Under the assumption that computers would soon be integrated into almost every aspect of professional life, the study concluded that there is a "national need to foster computer literacy" lest American science and industry fall behind the rest of the world. Individuals across the US workforce would face "unacceptable social and psychological costs" without such literacy; they could lose their jobs, find themselves unemployable, or see their workplace autonomy threatened in a changing economy that assumed a baseline of computer skills.[9] By the early 1980s, some of the study's predictions were being borne out in reporting about computers in the workplace. Features in national newspapers and magazines described executives who refused to acclimate themselves to computers despite acknowledging their benefits, resulting in added labor for the workers below them who were expected to master the new machines on their behalf.[10] Psychological studies similarly observed that many workers found the presence of computers in the workplace stressful. From a senior executive's perspective, they may have improved office communication and corporate decision making. However, these benefits were the result of extra work by clerical staff and junior executives, who felt that their jobs depended on their ability to develop new computer skills and that they were not rewarded for the extra labor that went into learning them.[11] Even among those who assumed computers would get easier to use, some suggested that in practice they would serve only "to further simplify and routinize work tasks and to reduce the opportunities for worker individuality and judgment."[12] In short, there was a growing belief that the pressure people were facing to quickly develop computer literacy was introducing new conflicts into the cultural and social contexts that computers were being integrated into.

Columnists writing in personal computing magazines took up these concerns, asserting that the present state of crisis facing personal computing was due to misplaced priorities in computer literacy training. In a series of articles published in *InfoWorld* beginning in late 1980, Ed Martino argued that a "crisis of epidemic proportions in computer literacy exists in this country. Very few business and professional people know enough about computers to cope with the emerging 'information age.'" The problem is that everyone assumed that "in order to be computer literate you must be able to program a computer competently in at least one high-level language."[13] Personal computers were still designed as if programming were their primary function. By 1981, however, the software industry was already a $150 million business. The market was roughly evenly split between first-party publishing programs, a smaller number of large third-party publishing companies, and a cottage industry comprised of hundreds of independent developers who sold most of their software directly to customers via mail order. In other words, by the early 1980s it was possible to use a computer without having to do much, if any, programming.[14] Describing his experience attending a consumer education program organized by Radio Shack, Martino noted that most people seemed interested during presentations that included software demonstrations but "tuned out" when programming tasks were discussed. The idea, he concluded, "that more than 5% of [new users] will ever write a significant application program is an elitist's delusion." Programming classes should still be available, as learning how computers "think" was a valuable skill, but it was not an essential one.[15]

Others argued that even if it were feasible to develop a pedagogy that would provide users with a detailed understanding of computation, rapidly changing technologies meant that their education would be obsolete within a few years. Schools and other sources for computer literacy education needed a model that focused on applications rather than on mastery of technical concepts. Lee The, associate editor for *Personal Computing*, explained that teaching novices, and especially younger students, a model of computer literacy based on programming or on the memorization of arcane command procedures would leave them "in the same position as a person who [was] taught blacksmithing in 1890."[16] We needed to stop thinking of computing as an end itself, he argued, and instead adopt a "computers-as-tools" mindset that teaches students that "what's most exciting about computers isn't the machines themselves, but what we can do with them." We are too busy teaching students about BASIC, he continued, when we could push them instead to consider the way that software could become a "creative writer's wings," a composer's "jazz band or string quartet," or a new sculpting medium. Those who supported a model of computer literacy focused on technical concepts, he noted, "argue[d] that programming is the most powerful (they love that word) thing we can do with computers," but what they failed to acknowledge is that it likely took them months if not years to develop their expertise. The computers-as-tools mindset emphasized that it was the user and not the designer who truly made a computer "useful."[17] To that end, The maintained that educators needed to develop a computer literacy curriculum that taught students and workers how to do what they wanted to do with these new machines rather than the technical skills and computational reasoning that computing professionals valued.

Many other journalists similarly suggested that a major source of the growing anxiety about computer literacy was the fact that the people designing computer systems did not try to understand the people who used them. Jim Edlin's columns for *InfoWorld*, often appearing alongside Martino's, argued that computer literacy was "both a hoax and a wasteful detour on the road toward the mass market micro." Personal computer manufacturers, he implied, had designed systems that put their needs first. Now that designers had figured out how to "build computers for the masses," the problem was to figure out how to "prepare the masses to cope with computers."[18] Some developers, he imagined, were "expressing the prejudices of their trade" by purposefully keeping computer systems complicated so that their own professional skills remained valuable.[19] Invoking a sense of exigency similar to that found in countercultural narratives of personal computing, he claimed that we should not "trust the computer professionals to lead us into the age of the mass-market micro" and suggested that rather than waste "valuable time and money on the chimera of computer literacy," we should adopt a radically different approach to software design, one that would make "efforts to the desperately important work of making computer-literacy unimportant. . . . The direct route to the age of the mass-market micro is not to make the masses computer-literate; it is to make the masses of computers human-literate."[20] In later columns, Edlin would elaborate on his conception of human-literate computing, suggesting that these machines should be ones that did not require documentation to use and ones that people would feel comfortable evaluating without the need for sales staff or spokespersons to explain their usability or usefulness to them.[21]

One reason why BASIC programming may have remained a focus of computer literacy education was that it was the one skillset that was generally recognized as transferable across hardware standards. Writing for *InfoWorld*, journalist Dorothy Heller illustrated how manufacturers tended to emphasize the differences between their products and those of their competitors. When she asked spokespersons from Atari, Apple, Commodore, Tandy, and Texas Instruments to define "user friendly," she noted that each company's response differed from that of the others in ways that appeared to highlight design features unique to its flagship systems.[22] In short, each wanted to give readers the impression that their systems were somehow unique and capable of providing benefits to users that could not be found elsewhere. There was some truth to this impression, but perhaps not in the way Heller's sources intended. All personal computers available to consumers during the early 1980s—whether general-purpose machines or the lower-priced appliance computers—included a BASIC interpreter as their default operating system. In theory, users could write a program in BASIC on one brand of computer and run it on another. However, due to differences in storage techniques, BASIC programs could often only be shared across different hardware standards if users typed their source code into the second computer. Additionally, differences between hardware components that affected how BASIC interpreters executed code often meant that a program would have to be modified after it was copied to another system in order to run correctly. These problems affected commercial software development too. Even if programmers released separate versions of an application for multiple types of personal computers, not all would share the same features, and most versions would look and feel different from one another during use.[23] In other words, users might have to relearn how to use a familiar application if they tried using a version written for a different hardware standard. Thus, while the ability to program in BASIC was generally understood to be a widely applicable skill in personal computing, there were significant limits to its usefulness. A shared hardware standard would make it easier for users and professional developers alike to transfer software and data across computers. It would also, in theory, make more room for training in general computing skills other than BASIC programming. Thus, many felt that a shared standard was necessary for personal computing to grow into something more than a hobby. IBM would essentially provide that standard, but it would not eliminate concerns about a looming computer literacy crisis.[24]

IBM Personal Computer

While IBM's development of its 5150 Personal Computer (PC) is not as well documented as that of the Apple II or the Macintosh, most accounts of the company's history agree on several major points.[25] The PC was first proposed by William Lowe in 1980. Lowe presented the idea of an open architecture product as a low-risk way for the company to explore retail computer sales. Shortly after getting approval, Lowe handed off responsibility for the project to Philip Estridge, who finalized the design and oversaw early production of the PC. The intentions behind the PC's design at times appear contradictory. As James Chposky and Ted Leonsis note, IBM had throughout most of its history attempted to limit the exposure of its research and development practices to the press. Much of what has been published about IBM's development process is the product either of a carefully controlled narrative produced by a handful of IBM executives, accounts from former employees, or speculation from professional "IBM watchers."[26] Similarly, Paul Carroll comments that his history of the company during the 1980s often relied more on conversations with people who worked for IBM's partners, like Microsoft, than on IBM's own employees.[27] Exploring these various accounts can help us to see how IBM developed and refined a rhetoric of usability across various stages of the PC's design, eventually representing its open architecture as a response to concerns about computer literacy. As I show in this section, IBM's executives and advertisements worked to reframe design features resulting from internal pressures and financial influences on the PC's development as if they were the product of a broad commitment to empowering users to develop the skills and knowledge necessary to integrate computers into their lives on their own terms.

Accounts of the PC's development emphasize two major influences: the success of the Apple II and the time constraints placed on the team by IBM's corporate management committee. Chposky and Leonsis allege that prior to his 1980 meeting with the corporate management committee, Lowe had studied the market carefully and concluded that Apple was "the leading producer of personal computers." One of Lowe's earliest proposals, in fact, was for IBM to buy the smaller company outright. Lowe felt that despite its initial success, Apple was "especially vulnerable" because it had made "critical mistakes" in "management, marketing, and research and development . . . that were never really rectified."[28] Specifically, Lowe believed that Apple's presentation of the Apple II as a hobbyist's machine was a mistake and that it needed to be marketed more aggressively to business users. In retrospect, historians have implicitly agreed with Lowe's assessment, citing VisiCalc as the "killer app" that helped lift Apple's sales as cheaper appliance computers grew popular. Yet VisiCalc's initial release on the Apple II is regarded more as the result of happenstance than any effort on Apple's part to cultivate business users. Most accounts of VisiCalc's development note that it was developed for the Apple II first only because its publisher, Dan Fylstra, preferred to work on a TRS-80 and decided to loan out his Apple II to VisiCalc's creators. Apple's executives also had reportedly expressed little initial interest in supporting the development of VisiCalc.[29] Nonetheless, Chposky and Leonsis note that Lowe associated the Apple II's potential for long-term success among business users with its open design. Because the Apple II's technical systems were well documented, he felt that there was and would continue to be greater availability of software for it compared to other computers available in the American market.

Some have suggested, however, that the influence of Apple's open design on the PC was more organic. For example, Carroll claims that prior to joining Estridge's team a significant number of the engineers who contributed to the PC's final design were devoted Apple II users. Many of those who were not reportedly purchased one shortly after joining the team and hearing about it from others.[30] Carroll even introduces Estridge in his book by describing him as "an Apple II devotee who loved to tinker with the one he had at home."[31] Estridge himself, however, has flatly denied any influence of the Apple II on the PC's design. When asked directly in a 1982 interview about whether his team had studied the Apple II, Estridge responded:

> No, we didn't. We didn't look closely at any single product. Instead, we looked closely at what purchasers were doing. We asked three kinds of questions: Why did the customer buy? What machine capabilities were the customers using? Why would people buy a personal computer in the future? If you hadn't purchased one yet, what were you waiting for? . . . We would certainly not call [the PC] a Super-Apple. We think there are a lot of features in the machine that stand on their own. It has some similarity to other machines but there are significant differences as well.[32]

Regardless of the precise nature of Apple's influence, most journalists noted that the PC's open design was a significant shift from IBM's previous leveraging of proprietary hardware and restrictive service contracts to maintain control over their machines after they were placed in a clients' offices or laboratories. Even those countercultural hobbyists who feature prominently in narratives that describe personal computing as a political movement in opposition to IBM were forced to admit that the company was "doing things our way" after getting their hands on a PC.[33]

The PC's open design also was influenced in large part by the time and budget constraints placed on its production. Despite being pitched as a low-risk way to explore retail computing, Lowe's proposal was approved with a short, twelve-month timeline for development. IBM's previous attempts to sell desktop-sized computers had taken much longer, and the machines had sold quite poorly. Before joining the PC design team, David J. Bradley had also worked on the design of IBM's Datamaster. Unlike the PC, the Datamaster had been designed more in line with IBM's past preference for proprietary components. Bradley notes that while development of the Datamaster began in 1978, it was released just prior to the PC in 1981 due to delays affecting almost every major component. According to Bradley, the PC team continually looked for ways to accelerate the design process to avoid problems like ones he had encountered while working on the Datamaster. The team ultimately concluded that the only way it could meet the timeline set by IBM's corporate management committee was to avoid using proprietary parts almost entirely. Bradley specifically points to the decision to license a BASIC interpreter and disk operating system from Microsoft, explaining that adapting IBM's proprietary interpreter for the Datamaster's processor had delayed the project by nearly a year.[34] Other engineers on the PC's design team have likewise explained that they first looked internally for hardware and software, accepting bids from existing divisions within IBM; however, they concluded that no existing division would be able to supply parts that met the timeline and price point approved by the corporate management committee.[35] Eventually, the design process reached a point after which IBM was essentially just licensing components from other manufacturers and rebranding them.

Bradley's account further suggests that IBM's management was willing to support an open standard because the company had tried and failed several times since 1975 to market microcomputers that relied on proprietary technologies. While the PC was briefly listed in catalogs with the model number 5150, the company denied shortly after its release that it was developed in connection with other products.[36] Nevertheless, the 5150 number connects it internally to a series of other desktop-sized computers that IBM developed in years prior and that were considered internally to have been commercial failures. While each of these computers sold for far more than the general purpose and appliance computers available in specialized computer shops and retail department stores, observers often described them as "personal computers." When IBM released the 5100 computer in 1975, for example, *BYTE* ran a story about it with the headline, "Welcome, IBM, to personal computing."[37] Prior to developing the PC, IBM also had produced three other machines in the 5100 series: the 5110 Model 1 (1978), the 5110 Model 2 (1978), and the 5110 Model 3 (1980), later renumbered as the 5120. In addition to the Datamaster, IBM also manufactured two other desktop computers that appeared on the market in 1980 that fit the definition of personal computer but that were not explicitly associated with the 5100 series: the 5250 and the DisplayWriter.[38] Although each of these machines may have superficially resembled a personal computer, they were priced and marketed by IBM as office automation systems. Each of these systems sold for anywhere between $9,000 and $20,000, depending on options, much more than any of the personal computers discussed in this book.[39] The inability of each of these machines to attract and sustain a userbase thus likely made the proposal for a more fiscally conservative attempt at personal computing amenable despite its requiring a break from IBM's past development practices.

Regardless of the precise nature or combination of influences that shaped the PC's design process, the result was a computer system that was far more open than any other, preassembled, non-kit-based computer. Prior to the PC, IBM had relied on third parties for at most only 10 to 20 percent of its hardware and software.[40] Yet with the PC, IBM not only bought off-the-shelf parts from other manufacturers but also published detailed technical specifications. As noted in the previous chapter, manufacturers of the cheaper appliance computers at times withheld information about their hardware and software in order to coerce third parties into lopsided publishing agreements and to create a more favorable market position for their first-party software and accessories. And although the Apple II also had a reputation for being an extremely open platform due to its own detailed technical reference, the machine still had several proprietary components that Apple was prepared to protect through legal action.[41] Fairly quickly after the PC first hit the market, other manufacturers, computer shops, and even some hobbyists quickly discovered that they could build a compatible machine by studying the technical reference manual and purchasing the exact same or similar parts, sometimes even relying on the same vendors that IBM had. Even those few components that IBM did develop itself, like its basic input output system software, were available from third-party companies that had been able to reverse engineer them.[42]

IBM's reliance on third-party components during the PC's design process also meant that the company introduced few new features. As a consequence, many PC reviews characterized the PC's design as unremarkable. Describing his first impression of the PC for *BYTE*, Gregg Williams said outright that the "genius of the people who designed the IBM microcomputer is that they managed to do everything conventionally but well—the IBM Personal Computer doesn't have any startling innovations, but it also lacks the moderate-to-fatal design problems that have plagued other microcomputers."[43] Will Fastie's review in *Creative Computing* similarly suggested that even though the PC's features were on paper unexciting IBM had managed to implement them in a way that had somehow exceeded his expectations. For example, he noted that the commands for its PC-DOS operating system were similar to those found in other disk operating systems but were better worded so as to more directly suggest their function and provided feedback that was "very clear" so that users would never "be at a loss for what to do next." The operations that were initiated by those commands were also fairly standard, but here they were performed with a "smoothness" not found on comparable systems.[44]

Comparatively lavish praise can be found in descriptions of the PC's documentation. Williams, for example, suggested that IBM's manuals would "set the standard for all microcomputer documentation in the future. Not only are they well packaged, well organized, and easy to understand, but they are also complete . . . [and] available much earlier in the life of this machine than it has been for other machines."[45] Like Williams, Fastie found the "the information content to be high, but clear. The writing is excellent. I have read the manual several times and have not been able to contrive a question for which the manual did not have the answer."[46] Even Edlin, who had previously railed against models of usability that relied on documentation as an inadequate response to the problem of computer literacy, praised IBM's manuals. Writing for *PC Magazine*, he explained that "IBM has designed a machine for the future. They have published a technical manual giving away in detail the secrets of their machine. And in that manual's pages one can read everywhere the deliberate effort IBM's designers have made to avoid hemming in the PC's future evolution."[47] No longer did Edlin believe that documentation would prevent personal computers from becoming usable. In fact, he suggested, the PC's might make them even more usable than ever. Many technology journalists and industry observers, in short, viewed the IBM PC as a step in the right direction in addressing the growing concerns about the relationship between computer literacy and usability.

It is worth pausing a moment to compare the PC's documentation with the Apple II's, which was similarly praised for its completeness, because it will help to illustrate some of the reasons that reviewers felt the PC's documentation would improve public opinion about the usability of personal computers. In 1978, Apple began including with its Apple II a volume titled the *Apple II Reference Manual*. While the manual did include some basic setup instructions and introductory BASIC programming lessons for novices, the depth of detail in its description of circuitry, hardware states, and machine code suggests that it was written for those who already possessed a technically sophisticated computer literacy.[48] In 1979, Apple began swapping the *Apple II Reference Manual* for *The Applesoft Tutorial*, a book presented as more appropriate for novice users. Yet, as an article comparing Apple's and IBM's documentation argued, *The Applesoft Tutorial* still "goes overboard . . . cramming its pages with hardware and software details that could clog the circuits of any beginner's brain."[49] Even though *The Applesoft Tutorial* included illustrations, used colored fonts, and was written in a more casual prose style than the *Apple II Reference Manual*, its pages were extremely text heavy. Its lengthy descriptions were likely intended to model a more conversational style; however, the writing often assumed a condescending tone implying that computing is still an imposition on everyday life. For example, the manual at various points referred to technical terms as the language of "computerniks," implying to readers that they were not a part of computing culture.[50] At times, the tutorial even apologized for restating information, noting in a patronizing tone that that "you hardly need to be told that anymore. In fact, you won't from now on."[51] Moreover, the bulk of *The Applesoft*

Tutorial was devoted to programming tutorials, and less than twenty pages were devoted to describing basic operation and computing concepts.

IBM's documentation for the PC was split into multiple volumes that were organized around a progression from novice to advanced use. When the PC was first sold to the public, the machine was packaged with the three-volume *Personal Computer Hardware Reference Library*, which included a guide to operations that served as a general reference and setup guide, a BASIC manual that offered instruction in programming concepts alongside a complete language reference, and a disk operating system manual that explained file system operations. Additionally, users could also purchase separately a technical reference manual, which provided detailed information about almost every aspect of the machine's hardware. While they are not formally numbered, each manual explains that readers should first review the contents of the guide to operations and then use the other volumes to follow-up on topics they would like to learn more about. The guide to operations, as its introduction noted, was intended to acquaint readers who were "new to computers" with foundational concepts and was "written in an easy-to-understand language."[52] Compared to Apple's documentation, it was much more concise in its descriptions. Instead of lengthy explanations, it made use of simple illustrations. These illustrations provided visual explanations to many general computational concepts, such as how floppy disks store data or the various outcomes of a COPY command (see figure 4.1). Although there was some overlap between the three manuals, reviewers noted that the introduction to BASIC and DOS found in the guide to operations was framed in a way that "assures that a novice can take advantage of the disk operating features and write simple programs using only the elementary manual."[53] The introduction to its chapter on BASIC, for instance, explicitly noted that its goal was to teach readers "enough BASIC" so that they "can use the SAMPLES program . . . and can run other BASIC programs" that they might purchase.

Thus, while the IBM PC did not establish a wholly new model of usability, the rhetoric of usability associated with it promised guidance and a clarity of explanation that many observers felt was not only unmatched by other manufacturers but necessary to support the kind of computer literacy needed to realize the widespread adoption of personal computers. Users would still need to do some studying to develop the literacy needed to realize the PC's full potential, but the included documentation would provide

- The file (LJCORS) is copied to another diskette and the name of the copy is changed:

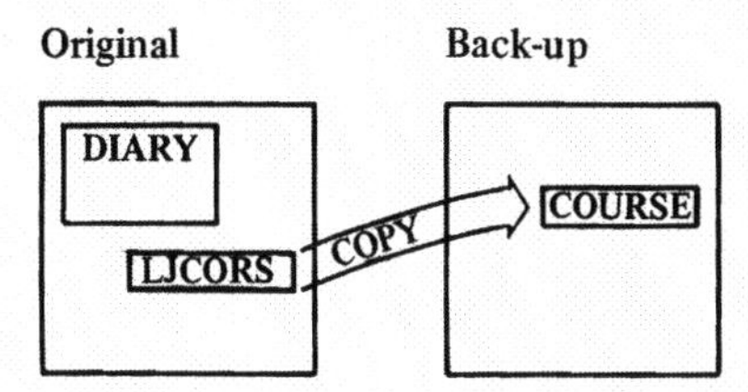

- The file (LJCORS) is copied onto the original diskette with a changed name:

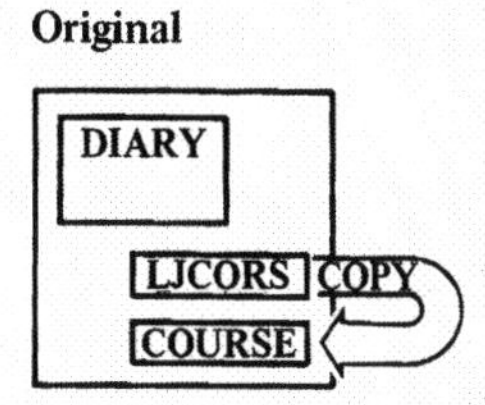

- The file (LJCORS) is copied onto the original diskette with a name that is already being used:

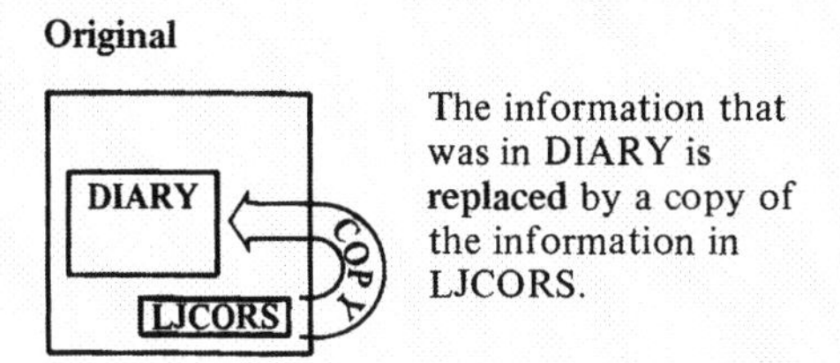

Figure 4.1. An example of IBM's use of descriptive images from its guide to operations booklet, included with the PC. Courtesy of International Business Machines Corporation, © International Business Machines Corporation.

users with a clear sense of direction throughout the process by delineating the different skills and understandings they needed to develop as they progressed.

The rhetoric of usability that stressed that computer literacy could be attained by anyone was not only apparent in IBM's documentation but also in its early advertisements." In its first ad campaign for the PC, "The IBM Personal Computer and Me," the company announced that its new machines were being built "just for you" so that you would "have a personal interest in" them (see figure 4.2). These advertisements tried to situate the PC in the home, showing individual members of a white middle-class family alongside text suggesting that each was not only able to learn how to use a PC but also capable of finding their own uses for one. Although the images and text varied, each similarly suggested that the computer and its documentation were written "in your language, not in 'computerese.' Our software involves you, the system interacts with you as if it was made to—and it was."

Advertising that promised software would be less confusing than a competitor's was nothing new, but what is important to note is that IBM tried to represent the PC as accommodating the distinct needs of its users rather than simply making it easier for users to adapt themselves to computing. IBM promised that its documentation offered a "structured learning process" that would get users started quickly. Rather than having to wade through the turgid prose of reference manuals for a few months to learn how to use the machine, "you can be running programs in just one day. Maybe even writing your own programs in a matter of weeks. . . . Once you start with it, you'll discover more than the answers and solutions you seek: you'll discover that getting there is half the fun." In short, IBM's advertisements for the PC sought to minimize any anxiety consumers might have about computer literacy by creating the perception that the PC's designers had prioritized universal usability. The PC might be a bit complicated to use at first, but IBM presented it as if its designers had tried to make its systems and documentation more human literate so that anyone could get the hang of using it pretty quickly.

Comments from executives echo many of these ideas, even going so far as to reframe the economic and organizational influences that produced its open design as resulting from a commitment to a universally inclusive model of usability. Until he was promoted out of the Entry Systems Division in 1984, Estridge served as the primary spokesperson for IBM regarding the

"My own IBM computer. Imagine that."

One nice thing about having your own IBM Personal Computer is that it's *yours*. For your business, your project, your department, your class, your family and, indeed, for yourself.

Of course, you might have thought owning a computer was too expensive. But now you can relax.

The IBM Personal Computer starts at less than $1,600† for a system that, with the addition of one simple device, hooks up to your home TV and uses your audio cassette recorder.

You might also have thought running a computer was too difficult. But you can relax again.

Getting started is easier than you might think, because IBM has structured the learning process for you. Our literature is in *your* language, not in "computerese." Our software *involves* you, the system *interacts* with you as if it was made to—and it was.

That's why you can be running programs in just one day. Maybe even writing your *own* programs in a matter of weeks.

For ease of use, flexibility and performance, no other personal computer offers as many advanced capabilities. (See the box.)

But what makes the IBM Personal Computer a truly useful tool are software programs selected by IBM's Personal Computer Software Publishing Department. You can have programs in business, professional, word processing, computer language, personal and entertainment categories.

You can see the system and the software in action at any ComputerLand® store or Sears Business Systems Center. Or try it out at one of our IBM Product Centers. The IBM Data Processing Division will serve those customers who want to purchase in quantity.

Your IBM Personal Computer. Once you start working with it, you'll discover more than the answers and solutions you seek: you'll discover that getting there is half the fun. Imagine that. IBM®

Figure 4.2. IBM PC advertisement printed in the January 1982 issue of *BYTE* magazine. Courtesy of International Business Machines Corporation, © International Business Machines Corporation.

PC. Generally, his comments mirror Lowe's idea that the PC was a low-risk way for IBM to test the retail computer sales market, reflecting IBM's tight control over how its research and development practices were represented to the public. For instance, when asked directly in a 1983 interview with *BYTE* why the PC adopted an open standard, Estridge echoed Lowe's framing of the decision as part of a conservative strategy developed through careful study of the personal computing market:

> We firmly believed that being different [by introducing a new standard] was the most incorrect thing we could do. We reached that conclusion because we thought personal computer usage would grow far beyond any bounds anybody could see in 1980. Our judgment was that no single supplier or single hardware add-on manufacturer could provide the totality of function that customers would want.[54]

Estridge also commonly characterized the PC's design as part of a broader strategy to solve problems related to compatibility and service in the growing retail environment. In this same interview, Estridge stated that they "knew dealers would have to provide warranty service" so they chose "commonly available parts . . . with the serviceman at the bench in mind. Our goal was to make the machine as easy for him to use as for a customer, because he's a customer too."[55] Thus, Estridge here characterized the PC's open design as an intentional move to take advantage of parts that had been "proven" on the market and that would facilitate coordination with IBM's retail partners at Sears Business Centers and ComputerLand rather than as an explicit commitment to supporting novice users.

Yet Estridge at other times moved outside of the IBM corporate framework and described the PC in terms of the idea that computers ought to be as usable as appliances. In a 1982 interview with *Personal Computing*, he avoided answering a question about how IBM views its competition and instead explained that

> knowing the business is not just knowing what the customer needs but knowing how possible it is for those needs to be satisfied by competitors. And surprisingly enough, at least in the personal computer part of the business, the competitor is not necessarily another personal computer. We are talking about the discretionary spending—the person may buy something else that is not a computer. It may be a high-tech item such as a smart TV set, a VCR, or even just going on vacation.[56]

Here, Estridge alluded to the persistent belief that novice users did not know how to evaluate the technical merits of a machine. What users wanted was the assurance that they would get a product that would be immediately usable by them and personally useful to them. However, unlike manufacturers of appliance computers, Estridge never portrayed users as disinterested in learning or unable to learn about technology. When asked by *BYTE* to confirm its impression that the PC was designed as a computer that would "make it easy for people to use—to experiment with the machine, to add to it," Estridge responded, "I think we're in an era in which the public has adopted personal computing in the same way it adopted the automobile. People want to know everything they can about it. That era will probably pass, but that curiosity is almost sensational now."[57] Far from being a drawback or a burden, the open but otherwise unremarkable design of the PC and its thick volumes of documentation were presented here as intended to support the ability of the average person to define their own relationship to computing.

Notably, Estridge semiregularly invoked ideas that recalled the hobbyists' model of personalization, stressing that individuals or communities of users would be able to configure the PC to suit their unique needs. In a 1982 interview with *PC Magazine*, for example, Estridge commented that he "believed we could build a machine that would be something special—so special that people who hadn't used IBM equipment before would use it."[58] While Estridge hoped that the PC would be useful also to those already familiar with IBM's products, he often appeared to recognize that the identities and needs of novice users were more varied than those with a technical background like himself. Elsewhere, when asked directly to describe who IBM envisioned as its ideal user, he alluded to the idea of conviviality:

> I don't think we have a typical user because the machine is so communal that typical doesn't have meaning, except for the fact that more and more people are discovering that they have needs that can be answered rather nicely by a personal computer. And they are in all walks of life—all the way from very young children to very elderly people—in every profession.[59]

Reflecting back on the initial commercial success of the PC in a 1984 essay published in *Creative Computing*, Estridge reiterated his belief that an open design was necessary to support the diversity of interests, backgrounds, and identities among novice users: "there is no 'one size fits all' in personal computing. Each person is unique, and has different needs, habits, incomes and desires." Giving users more choices was key to acknowledging the heterogeneity of users, and choice was "the underpinning of IBM's commitment to open architecture."[60] In moments like these, Estridge reframed a design resulting from a hurried process by a company with little experience selling directly to consumers as intended to support a model of usability that linked personalization and universal use. Given access to technical knowledge and a structured program of learning, anyone could develop a degree of computer literacy that would allow them to make their own choices about how to integrate computation into their lives. And per Estridge, IBM was committed to ensuring users had their freedom.

Nonetheless, by the mid-1980s IBM's rhetoric of usability would begin to shift more toward the practical benefits, immediacy of use, and simplicity of operation stressed by manufacturers of appliance computers. IBM ended its "IBM and Me" campaign and began its "Modern Times" series in mid-1982. Chposky and Leonsis claim that IBM had always planned to phase out its early advertisements and to replace them with the more well-remembered series featuring Charlie Chaplin's Tramp as part of an effort to soften its corporate image.[61] This second series roughly alluded to Chaplin's critique of the dehumanizing effects of industrialization—a wink at the company's own past reputation—suggesting that the PC represented a new era of wonder and a legitimate promise of empowerment through technology. When one looks closely at the rhetoric of usability developed across these new advertisements, it becomes clear that the changeover from celebrating computer literacy to emphasizing an appliance-like immediate utility was gradual. Initially, the new advertisements echoed earlier ideas, highlighting flexibility and customization of the system while also touting documentation that would "teach you to use it with the greatest of ease" so that it becomes "your tool."[62] Carrying forward the assumption that its users were learning to program, the "Modern Times" series even began in early 1983 to invite users to submit their own software for possible publication through IBM.[63] But by the end of 1983, the advertisements began to shift almost entirely to promoting specific tasks, like connecting the PC to information services or particular software packages available through IBM's first-party publishing program.[64] This change not only illustrates the growing influence of the rhetorics of usability associated with appliance computing but also of the by now long-standing skepticism about the value of consumers directly engaging with computational concepts.

While few systems manufacturers copied IBM's rhetoric of usability closely, many did reproduce its model of usability. One unintended consequence of using commonly available components was that smaller companies were able to produce PC-compatible clones legally.[65] By 1982, established electronics manufacturers like Hewlett-Packard and new ones like Compaq had begun selling computers modeled after the PC's design that could use most of the same software. Reflecting in 1984 on the state of the market, journalists Charles Rubin and Kevin Sterhlo explained that IBM's PC had

> stimulated the rapid spread of [personal computers] to the desktops of corporate America. To be sure, thousands of individual managers had been using Apples, Radio Shacks, and other brands, but IBM's entry convinced corporate purchasers that personal computing was here to stay, and that it was no longer a gamble to take advantage of the new technology. With the acceptance of personal computers by large institutions, a much larger percentage of individual purchasers now felt comfortable. Demand for the IBM Personal Computer was so great that at one point there was a backlog of orders several months long.[66]

Rubin and Sterhlo estimated that by 1984 as much as 85 percent of personal computer software available for sale had been developed for use on PC-standard hardware. According to them, the rapid expansion of the personal computing hardware and software markets owed to IBM having brought a level of seriousness to personal computing that earlier start-ups and appliance manufacturers did not. Additionally, because the PC's open design allowed it to serve as a standard platform for software development, the greater availability of software helped to foster an increased perception of the usefulness of personal computers. As I discuss in the next section, however, IBM's influence on the American personal computing market did not change the public perception that too sophisticated a degree of computer literacy remained necessary to use these machines comfortably.

Criticism of the IBM Model

Even as journalists credited IBM's PC with conferring a sense of legitimacy onto personal computing in the early 1980s, it is important to understand the scale of adoption. In 1981, approximately eight hundred thousand personal computers were sold in the United States, but starting the following year numbers would more than double, reaching approximately 2.5 million in 1982 and 6.2 million in 1983 before beginning to plateau in 1984 at 7.8 million.[67] According to the United States Census Bureau, there were approximately 283 million people living in the country in 1980.[68] While the growth of annual sales of personal computers into the millions could give the impression that the market was far reaching, research firms estimated that only 2 percent of homes and 5 percent of businesses owned one. Citing these numbers, technology journalist Peggy Zientara argued in 1983 that her colleagues needed to recognize that the adoption of personal computing was not as widespread as it was often portrayed in American computing magazines.[69]

In other words, many industry observers remained skeptical of claims that personal computing was becoming commonplace. Writing in *BYTE*, Sam Edwards encouraged his readers to look back through the advertisements scattered through its pages and ask whether or not any products they used actually lived up to the experiences their developers claimed: "The 'ease of use' idea has caught the software industry's attention, but so far it has generated mainly advertising copy. Rare indeed is the advertisement that doesn't proclaim its program Easy to Use or User Friendly. Investigate these claims and you will usually discover just another hard-to-use product with a bunch of lessons and help screens tacked on."[70] Edwards even calls out IBM by name, noting that the assurances in the company's rhetoric of usability that users could learn to handle complex technical information had lead some software developers to "equate difficulty with sophistication."[71] From this perspective, IBM's model of usability had not actually addressed the problems its rhetoric claimed to and had instead only made things worse by encouraging other developers to inundate users with information that for novices likely seemed, at best, only arbitrarily related to the tasks they were hoping to perform. Rubin similarly suggested that IBM's rhetoric gave people the mistaken impression that they "could figure the thing out on coffee breaks or between appointments" when in fact learning to use a computer was "more like taking up a musical instrument."[72] Jeffrey Rothfeder further argued that personal computer interfaces had not fundamentally changed since the mid-to-late 1970s; "easy computers," he noted, should not require users to "learn how to communicate with the machine anymore using its terms" and instead should be "commanded to communicate on people's terms."[73] Computer literacy was still a significant concern. But as comments like these suggest, the problem was not with the users who tried to learn, the companies that were integrating computers into their work processes, or the public and private efforts to teach people about computers. The problem was that the status quo of design favored models of usability that imposed a seemingly undue burden on users.

Some journalists even asserted that a conflict between computation and culture was the source of lingering concerns over computer literacy. Designers who followed the IBM model appeared to be looking for ways to push computational reasoning into culture rather than find ways to integrate computers seamlessly into it in a more subtle way. So long as computers required what many felt was an advanced degree of computer literacy to be seen as truly usable and useful, they would continue to be seen as interfering with the purposes and contexts they were ostensibly designed to support. Paul Heckel, for example, discussed this idea regularly in his column in *InfoWorld* called Designing for People. He repeatedly tried to persuade software developers to adopt an editorial mindset:

> George Orwell said that "good prose is like a windowpane." The best writing is so transparent that the reader does not see it as being there. The thoughts come through unclouded. We can use the user's work as the source of interest. But to do this well, we must make the user interface transparent. The user sees himself as doing his work and does not see himself as using the computer. Software must not get in the user's way.[74]

In other words, the more information offered to users, the more time they had to spend wholly redefining a task or topic they already believed they understood in terms of the new, unfamiliar concepts presented to them by software. To keep users from being overwhelmed by too much information, software developers needed to edit their interfaces so that they presented a readily identifiable framework for action, ideally one that drew on a rhetoric of usability which could describe that framework in terms that users already understood. "Friendly" software, he often asserted, is software that does not leave users constantly second guessing themselves.

Thus, a consensus was beginning to form among technology journalists and software developers that there needed to be a way to represent computing without reference to computation. Doing so would narrow the range of skills and understandings that users could develop, but it would also reduce the number of technical concepts they had to master to experience computers as personally usable and useful. In the course of critiquing IBM's model and rhetoric of usability, Edwards explained how he had been trying to realize this form of user-friendly design within his own company. The next generation of software must "offer [users] less. The thinner the manual, the less you'll have to read to learn the program. The fewer the choices on the menus and prompt lines, the less you'll have to think about before making your choices. The less a program does, the fewer things can go wrong with it."[75] When using a computer, the user's focus should always be on *doing things* with the computer and not on *managing computational processes*. As others began to espouse views similar to Edwards's, they developed a new narrative framing for the personal computer revolution that celebrated a lack of computer literacy. Paul Kellam, for example, wrote in an editorial for *Personal Computing* that soon innovations in interface design would move "us towards a future in which computing will be easier than driving an automobile. . . . And in the new generation of personal computing, you won't have to know much about how computing works in order to compute."[76] Although a number of hardware and software developers were looking to create such an experience, Apple would become the first company to release a commercially successful "transparent" computer. In doing so, Apple positioned its new products as a response to the emerging consensus that computers needed to be immediately and universally useful. Apple would develop a rhetoric of usability for the Macintosh that similarly reframed the idea of mass computer literacy as oppressive and offered in its place a rhetoric and model structured around interfaces that represented computers in noncomputational terms.

Apple Macintosh

Dramatizations of the Macintosh's development process often frame its transparent design as the product of Steve Jobs's singular vision for computing or of Apple's success in commercializing a technology that it "pirated" from Xerox.[77] While there is some truth to these sensationalized depictions of the Macintosh's design process, both speak to the way that Apple's approach to usability has been mythologized in American popular culture as a critique of a stubbornly monolithic computer industry run by out-of-touch engineers. Although Apple has always portrayed itself as challenging the status quo of professional engineering culture, by the 1980s its corporate structure had come to resemble that of most other large electronics manufacturers in the United States.[78] All of these aspects of the Macintosh's history and its design process are well documented. My goal in this section, however, is to set aside that mythology and engage with two often unacknowledged influences on the Macintosh's representation of user-friendliness: the appliance computers and office automation systems of the 1970s. If we resist the tendency to conflate the origins of transparency with graphic user interfaces (GUI), then we can begin to see that the Macintosh's rhetoric and model of usability have influences that extend beyond Jobs's direction to copy Xerox's technology. Through the Macintosh, Apple promoted a seemingly new approach to software design tailored to "knowledge workers"; however, as I show in this section, many principles associated with the success of the Macintosh's design and presentation had already seen a degree of commercial profitability during the previous decade. The models of usability associated with appliance computing and office automation were generally considered to be at odds with the vaguely democratic implications of open architecture personal computer systems. But Apple would reframe its closed architecture as liberating through a concerted effort to develop a counter rhetoric of usability that contrasted with IBM's vision of personalness.

When it was revealed to the public in late 1983, the Macintosh was presented as a machine that embodied a wholly new approach to computing, one that was developed specifically to challenge an older usability paradigm now dominated by the IBM PC. However, many elements of the Macintosh's rhetoric and model of usability were originally developed in response to older machines and defined before the PC project had even been proposed internally within IBM. The Macintosh project was begun in May 1979 by Jef Raskin, a technical writer and Apple's director of publications, who described the ideal user of his machine as the "person in the street": someone who enjoys a "perverse delight" in saying "I don't know the first thing about computers." Raskin outlined a number of characteristics that the Macintosh could not have if it were to appeal to this type of user before stating in summary that "any system which requires a user to ever see the interior, for any reason, does not meet these requirements."[79] He also argued that it must have a very specific style of interface that must not under any circumstances introduce too much technical information: "Computerese is taboo. Large manuals, or many of them (large manuals are a sure sign of bad design) is taboo. Self-instructional programs are NOT taboo."[80] Given that these comments date to 1979, Raskin was not contrasting his proposed computer with the IBM PC. While one could read his comments, like those from

Tandy's and Commodore's spokespersons, as criticisms of the kit-based machines of the mid-1970s, it is important to recognize that after 1977 manufacturers like Altair, Cromemco, IMSAI, SOL, and Ohio Scientific had begun to restructure their business models around providing custom systems to data and technology companies.[81] It is more likely that Raskin's criticisms were here directed at the Apple II, a machine for which he himself had been working to develop robust documentation.

The Macintosh's hardware as described by Raskin shared many features with the appliance computers of the late 1970s. Just as Chuck Peddle insisted that users could take their PET to a local appliance shop for service, so Raskin pressed the point that users should have no need to access the machine's internals at all: "If the computer must be opened for any reason other than repair (for which our prospective user must be assumed incompetent) at the dealer's, then it does not meet our requirements."[82] A closed design would require that all Macintoshes have an identical hardware and software profile. Yet to maintain parity with other appliance systems, Raskin did concede that the Macintosh should support some minimal number of external components. He established two conditions for any such peripherals. First, each would need to "stand on a table by itself," "have its own case," and "look like a complete consumer item in and of itself." They should require no more knowledge on the part of a user other than how to plug them into the main unit. Second, Raskin argued that the Macintosh "must be in one lump" so it would have a similar profile to that of the PET.[83] Any hardware peripherals that would be required for operation would need to be incorporated into the unit's casing. Although this emphasis on providing consumers with a complete experience out of the box was common to the rhetoric of appliance computing, it was a distinctly different approach to usability than that of the Apple II, which offered users a distillation of the hobbyists' philosophy of free access to computational systems.

As Raskin continued to develop his ideas for the Macintosh, he began to more directly compare the project to the appliance computers available during the late 1970s. Price, he explained, would be a key deciding factor, perhaps the most important one, as "random consumers . . . do not know enough to go much beyond the advertisements and the bottom line."[84] These remarks are similar to those documented in the previous chapter from executives at Tandy, Commodore, Atari, and Texas Instruments, who each in different ways stressed that novice users likely would not be able to evaluate any computer on its technical merits. Although Raskin believed that Apple was capable of engineering an appliance computer that was of far higher quality than "the $795 Pet, the $698 TRS-80, the Atari 400, and some other machines now coming along, it is clear," he explained, that Apple needed "a product that looks (and is) competitive." The only reason Tandy had outsold Apple during the late 1970s, he argued, was because its hardware was cheaper: "It is hard to see how anybody could put up with a TRS-80 if they had had much experience with an Apple II," but "the fact is that a random customer will not have the opportunity to make the comparison beyond seeing the [price] difference." The Macintosh should above all else thus be "an Apple quality product for a low entrance price."[85] While Raskin may have had criticisms of the Apple II, he is very clear here that the Macintosh would not compete against it. His proposal was driven by the argument that Apple needed to make a move into the fast-growing market for entry-level, appliance computers.

Raskin modeled many features of the Macintosh after computers by Tandy, Commodore, and Atari, but he did make one significant departure. For Raskin, an appliance computer was not meant for programming. As I have noted, programming was still considered a primary activity in personal computing almost a decade after the availability of the first computer kits in late 1974. Raskin acknowledged as much in 1980 by commenting that the conventional view had always been that "to get the full benefit from a computer and to exploit its inherent flexibility the user must program it in some form." However, the main reason why personal computing outside of the workplace remained a niche hobby was that "programming, as a human activity, rates with torture in the popularity polls."[86] While he expected that the market would soon grow to a point where users would not have to write much, if any, of their own software, he expressed a concern about the usability of commercial applications. In his experience as a technical writer, software written by professional programmers "more commonly reflect[ed] the difficulties encountered in programming than the real needs of the user."[87] A true appliance computer would not only provide all of the functions consumers might require without the need for additional software purchases or upgrades but would also be something for which they could immediately identify clear uses. For Raskin, the idea of a "personal computer" did not refer to any sort of individual or unique relationship to computation that users might define through creative exploration of technology; rather, a personal computer ought to have an "applicability to work, homelife and play; portability or wide geographic access or both; its importance or significance to the user in all these roles; affordability; and options to make the technology aesthetically acceptable to the owner."[88] In short, a technology only becomes "personal" if someone finds a use for it. If the Macintosh supported a small range of specific tasks and did so well, then it was more likely to be seen as personally useful than other appliance computers, regardless of any limitations resulting from its lack of programmability.

Raskin's papers were not the only source of early influence on the Macintosh's rhetoric and model of usability. Technologically, the Macintosh owes much of its design to the Lisa, a computer that Apple specifically marketed as intended for knowledge workers. Development of the Lisa began in 1978 as a next generation Apple II: a fairly "conventional" $2,000 computer but the first from Apple to incorporate a 16-bit processor.[89] It was not originally proposed as a GUI-based system. By the 1980s, however, the Lisa that was shown to the public looked much different and was described in terms not often associated with Apple's typically hobbyist-inspired marketing copy. While Apple insisted in some of its advertisements that the Lisa was a "personal computer," journalists were quick to observe that at a cost of $10,000 it was in a distinctly different class of hardware.[90] Writing in *Personal Computing*, Michael Rogers described the Lisa as part of plan to "replace everything in [the] office" with Apple products without "chang[ing] the way people did things." The profile directly identifies the Lisa as a product separate from Apple's personal computers, an attempt "to do for office systems what the original Apple did for the entire field of personal computing."[91] Here, Rogers noted that the Lisa had less in common with the new kinds of computers that were being marketed directly to consumers and more in common with the high powered—and high priced—office automation systems that had been marketed to businesses since the 1960s.[92] Most accounts of the influence that Lisa had on the Macintosh emphasize how the Macintosh refined the Lisa's GUI, streamlining its operations so that they could be ported to the Macintosh's more affordable hardware. Yet the influence of the Lisa on the Macintosh was not merely technical. The Macintosh's rhetoric and model of usability owe much to the Lisa's positioning as an office automation system.

Office automation systems had been promoted since the 1960s as user-friendly computers that could be used right away with very little training and without requiring users to directly engage with their more technical aspects. Major office automation manufacturers included IBM and DEC; however, a smaller company, Wang Laboratories, most consistently incorporated the kind of language into its promotional materials that would later both be associated with transparent design and be emphasized in the rhetoric of usability surrounding the Macintosh. As early as 1977, Wang was promoting its office computers as ones that "everyone can use": "The PCS-II is people-oriented, it is a tool for all."[93]

Brochures from the mid-1970s did note that users would need some minimal training but explained that Wang would be happy to provide it. Later brochures began to incorporate the idea of a computer literacy that was not tied to a specialized computational reasoning by promising that Wang's computers could be used without training or manuals.[94] Whereas some of Wang's early promotional material discussed how the systems were programmable and adaptable to a variety of tasks, language referring to the technical aspects of computation had largely disappeared by the 1980s. A brochure from 1981 even claimed that Wang's computers were "as easy to use as a typewriter, even for people who've never seen a computer before."[95] Wang's computers were presented as tools that were universally usable as a consequence of their design being focused wholly around specific purposes like word processing, account management, and data processing. Wang's customers were not expected to customize or otherwise concern themselves with computation. In this sense, office automation systems like Wang's shared many aspects of their rhetorics and models of usability with those of appliance computers.

Although none of the computer systems Wang sold implemented a GUI, a competing office automation system being developed contemporaneously at Xerox's Palo Alto Research Center did. By most accounts, Lisa's design was revised to incorporate a GUI after Jobs and some of Apple's engineers were shown demonstrations of the interfaces that Xerox was developing for its Alto computer and its SmallTalk programming environment. Lori Emerson has documented how many of the same lessons that Apple took away from the demonstration were ones that Xerox would implement in its own attempts to commercialize its GUIs. SmallTalk was produced as part of Alan Kay's efforts to design technologies that would make computation more legible to novices.[96] It introduced an interface that represented software and data as "objects" that users could manipulate directly. This interface was intended to help users conceptualize relationships between software and data in such a way that they would not have to describe them anew each time they wrote out a series of commands. Yet the Alto was never manufactured widely and SmallTalk remained a software development tool. Xerox instead moved to commercialize the "object-oriented" approach of SmallTalk via the interface for its Star computer system, a product intended explicitly "to make 'knowledge' workers more productive."[97] Emerson describes the Star's interface as "conflicted at its core," an attempt to merge Kay's ideas with office automation concepts. Whereas visualization could offer users new ways to organize technical information, metaphorical representations could promote specific interpretations of software and data linked to the tasks that users were performing rather than represent the computational processes that supported them. The resulting model of usability was one that provided users with "highly restrictive commands in the name of simplicity (restrictions that certainly excluded certain creative possibilities)," supported by a rhetoric of usability that fostered only "a superficial understanding of the system."[98] Although the Star's development occurred largely in parallel to the Lisa's and the Macintosh's, Emerson's work shows that both Apple and Xerox recognized that while GUIs could serve to make complex computational concepts more legible to novices, they could also be leveraged to divorce representations of computer-assisted activities from the technical concerns of computation.

Apple would incorporate ideas similar to those associated with Wang's and Xerox's office automation systems in its advertisements for the Lisa. A lengthy magazine insert, for example, explained that with "a conventional personal computer, you first have to program yourself by studying the manual and learning a complex set of computer commands that vary widely from program to program." The Lisa, on the other hand, is designed so that "you can work with the system intuitively, right from the start. . . . You can concentrate your effort on what you want done—not on how to get the computer to do it."[99] One key difference between the promises that companies like Wang, IBM, and DEC made about the usability of their office automation products and those made here by Apple about the Lisa was that the former companies had dedicated technical staff that would setup, maintain, and develop software for their clients. In pushing for a personal-computer-like experience, Apple promised its users that Lisa would be a standalone system providing the same benefits. Some industry observers, however, were skeptical that Apple could deliver the same kind of ease of use without the same support services that Wang and other office automation companies offered.[100]

Apple also incorporated user-friendly concepts similar to those promoted by Wang into the Lisa's model of usability by developing two separate interfaces, one for office workers and one for software developers. Apple did not coin the term "knowledge worker," but its advertisements suggest that it understood the group to include "executives, managers, and small business owners." When discussing the Lisa, many journalists compared it to computers designed for interactive data processing in corporate environments like those used by large corporate banks and financial analysts.[101] In short, the Apple intended the Lisa to be used by people who could benefit from the programmability of a computer but whose time was viewed as too valuable for them to be doing the programming themselves.[102] Thus, the Lisa's default user environment did not include access to programming tools. Instead, the Lisa booted to a desktop environment that provided access to a set of software applications that resembles today's productivity suites: a word processor (LisaWrite), a spreadsheet (LisaCalc), a database manager (LisaList), a graphic design tool (LisaDraw), and a project management tracker (LisaProject). If a Lisa had the "workshop" package installed, then users could select at start-up to enter into a minimalist command-line environment that freed up system resources normally dedicated to supporting the GUI. This separate interface provided access to programming tools, initially only in the Pascal language.[103] In short, the Lisa's model of usability itself enacted a separation of computer literacy from computational literacy.

While reviewers recognized and accepted the Lisa's model of usability as appropriate for office automation systems, many argued that those features the Macintosh shared with it were inappropriate in the context of personal computing. As John Anderson noted in his review for *Creative Computing*, the Macintosh's design broke from the broadly shared expectation that personal computers should support free access to computation. Despite the increasing popularity of the IBM PC standard, the Apple II continued to be an industry-leading system. In fact, he suggested, its open approach was the main reason Apple had remained influential while other early manufacturers from the 1970s had by now folded: "By precluding easy hardware expansion on the Mac, Apple writes off a major component of its early success [with the Apple II]: expansion flexibility. Sure, it might have taken some imagination at first to envision the kinds of cards the Mac might need. But if an expansion bus were available, people would start to invent them." Instead, Anderson concluded, Apple presented its new computer as if it "already ha[d] everything you need."[104] Most reviewers suggested otherwise, arguing that the Macintosh prioritized a flashy interface over supporting basic features that many users had come to see as necessities during the early 1980s. One particular sticking point that many reviewers noted was that the Macintosh's lack of internal expansion initially limited users to a single disk-drive. Apple sold external disk drives, but these would not be available to consumers until several months after launch. While personal computer software was written so that no more a single drive was necessary at any given moment, it was very common for users to install a second drive so that they would not have to eject disks frequently. Gregg Williams, *BYTE*'s then editor, even broke from the neutral tone the magazine tried to effect in its reviews to vent his frustration: "I am not alone in this feeling this way; the first thing two *BYTE* editors said when they saw the first Macintosh was, 'Only one disk drive? You've got to be kidding!' After numerous disk swaps trying to load Mac Paint from one disk and a drawing from another, I am convinced most users will eventually buy the second disk drive."[105] The use of one drive for application software and another for data storage was common enough by the early 1980s that most personal computers were sold with two. Williams was also concerned about the lack of programming tools, as "no one has sold a computer without [built-in support for] BASIC (or some other language) in years."[106] Williams viewed these omissions as part of a similarly predatory strategy that manufacturers of cheaper appliance computers had relied on. This strategy was evidently not one he expected to find in a computer that sold for $2,500. In pointing out these limitations, reviewers asked readers to consider whether or not the allure of its GUI was worth the reduced functionally. If the ease of use it promised was not a primary draw, then "the Mac is reduced to a rather average machine."[107]

For its part, Apple—and especially Jobs—invited comparisons between the Macintosh and other computers, particularly the IBM PC. In addition to producing the *1984*-inspired commercial, Apple took out full-page advertisements in newspapers taunting IBM, and Jobs frequently attacked his company's competitor in internal presentations to stockholders.[108] Yet Jobs's interviews promoting the Macintosh reveal much more about the company's internal politics and the motivations behind the machine's model of usability than they do about the company's view of the IBM PC. When contrasting the Macintosh to the IBM PC, Jobs often alluded to conversations among industry observers, like those I have described in this chapter, who felt that the next generation of computers would be ones that users did not need to think about. In a 1984 interview, Jobs refers to the PC as being built around "1970 software."[109] By contrast, the Macintosh's design was pushing computing toward a "point where the operating system is totally transparent. When you use a Lisa or a Macintosh, there is no such thing as an operating system. You never interact with it; you don't know about it." Users, he explained, were "much more concerned about what the computer will do" rather than how it worked, which was the "right way of thinking about products."[110]

By pointing at the PC, however, Jobs also implicated the Apple II. Reading these remarks along with those he is alleged to have made privately to other Apple employees reveals a different set of motivations driving the Macintosh's model of usability. During the early 1980s, Jobs reportedly had become frustrated by the fact that third parties had more control over how consumers viewed the usability and usefulness of the Apple II: "We don't have control, and look at all these crazy things people are trying to do to it. That's a mistake I'll never make again."[111] The remarks show that Jobs shared Raskin's goal of implementing an appliance computer model of usability for the Macintosh. But whereas Raskin appeared to be interested in this idea because he felt confident that Apple could realize the ease of use promised in the rhetoric surrounding appliance computing, Jobs's interest seemed to lie more in the business strategies that their models of usability made possible. Moreover, Jobs also seemed to have adopted the hobbyists' revised framing of the personal computer revolution as something that could only be realized by elite technologists curating computation for the general public: "Customers don't know what they want until we've shown them."[112] In this view, average consumers did not even possess the computer literacy necessary to explain how or why they wanted to use a computer, let alone actually realize those goals when one was placed before them. Maintaining tight control over the hardware and software ecology that would grow around the Macintosh was not just good business but essential to Apple's ability to support a definition of computer literacy divorced from computation.

While Apple would promote the Macintosh as being usable and useful for a variety of creative ends, Jobs often adopted a rhetoric of usability that was associated more with office automation than with personal computing to explain the appeal of the Macintosh's model of usability. When discussing how the Macintosh would simplify problems of computer literacy, he explained that "it takes 40 to 110 hours to learn to use an Apple II. That may be acceptable to a spreadsheet junkie, but to a person who is going to be using a personal computer half an hour a day, or maybe an hour a day, you will not get that person to spend 40 to 100 hours learning how to use a computer."[113] A more appropriate goal for developing a minimal degree of competency, Jobs noted, should be twenty minutes. Jobs here suggested that there was a tension between computation and the cultural and social contexts of computing. Personal computers—and by extension their designers—demanded things of us in exchange for their promised benefits. The Macintosh could resolve this tension, but only if we stopped assuming that computational literacy had any value outside of highly technical contexts. Although the Macintosh would limit the ability of users to develop computational skills and acquire computational knowledge, in the end, it would expand the range of things users could do with personal computers. By analogy, Jobs suggested that calculators did not bring about the end of mathematics. Most people have "never learned how to use a slide rule," yet "almost everyone knows how to use a calculator."[114] The move from slide rules to programmable calculators only helped people to more effortlessly incorporate mathematical reasoning into more of the decisions they made. Like the journalists I have cited in this chapter, Jobs argued that redefining our computer literacy so that it did not rely on an understanding of computational concepts would allow people to more immediately recognize the usableness and usefulness of personal computers. If we changed our expectations about what counted as computer literacy, then it really could be something acquired over a cup of coffee, incidentally rather than through self-study.

A mixture of appliance computing and office automation rhetoric can also be found in Apple's advertisements for the Macintosh, expressed through a countercultural affect that framed computation as oppressive and the company's new, narrower definition of computer literacy as revolutionary. In December 1983, Apple began promotion of the Macintosh through a nineteen-page brochure included in several national news magazines that described it as a computer "for the rest of us."[115] The brochure also at times directly engaged with the language IBM used to promote the PC, offering a counter rhetoric of usability. Whereas IBM had suggested that the PC standard had made personal computing more accessible than ever, Apple's brochure instead suggested that personal computers were still a rare sight: "In the olden days, before 1984, not very many people used computers, for a very good reason. Not very many people knew how. And not very many people wanted to learn." If IBM described a public eager to find new ways to use the PC, Apple portrayed its interest as faltering: "In those days, it meant listening to your stomach growl through computer seminars. Falling asleep over computer manuals. And staying awake nights to memorize commands so complicated, you'd have to be a computer to understand them." Completely abandoning the idea that computer literacy could lead to personal empowerment, Apple instead asked its readers to consider a seemingly simple proposition: "Since computers are so smart, wouldn't it make more sense to teach computers about people, instead of teaching people about computers?" Designing easy-to-use computers in a language that "humans could understand" would mean abandoning rhetorics and models of usability that were based on overly technical concepts. Apple thus asserted that its innovative approach to design had finally, and simply, solved the complex problem of computer literacy. Developing a language of computing that emphasized skills and understandings that users already possessed would resolve the tension between computation and culture because it would prevent merely technical concerns from drawing our attention away from the things that really mattered to us.

The remainder of the brochure focused on how the Macintosh's model of usability supported those things that really matter. While its visual interface seemed to call attention to itself, the brochure explained that the Macintosh would be both phenomenologically transparent and transparent to human intention. Users would not have to think about the Macintosh during use, and its simple interface would not require them to reconceptualize the activities they were hoping to perform with it. In a series of juxtaposed images of computer screens, Apple illustrated that the "difference between the Macintosh and the PC becomes obvious the minute you turn it on." By comparison to the PC, the Macintosh "seems extraordinarily simple . . . because conventional computers are extraordinary complicated." The brochure stressed the way that the PC's interface made demands on users to shape their tasks to the system. Apple's computer did not make such demands: "If you can point, you can use a Macintosh." Users would be limited only by their imagination, "not the limitations of the computer."

The Macintosh naturally supported users and would not interfere or influence their intellectual activity: "If you can't make your point with a Macintosh, you may not have a point to make." For the personal computer revolution to be truly, finally realized, the brochure suggested that people needed to be able to access computers free from the influence of computation. As Apple would say in its advertising campaigns over a decade later, it was time for us to "think different" about computer literacy—which is to say, not like the IBMs of the world. Ideally, we would not concern ourselves with it at all. Computation was remote from our lived experience, and its demands oppressive. Rather than worry about whether your computer literacy was sophisticated enough to understand computation, you should leave it to the professionals. You had more important things to do.

Despite any reservations journalists may have had about the Macintosh's model of usability, they were largely accepting of the rhetoric surrounding it. Some even lavished so much praise that they sounded like they were part of Apple's public relations team. Rubin's review in *Personal Computing*, for example, opened with prose reminiscent of the kind of speeches that Jobs would deliver when introducing the Macintosh to Apple shareholders:

> Nine years ago, Apple Computer issued a challenge to the world. With its personal computer concept, Apple challenged us to change our attitudes about what computers were, and how we used them. Millions of us took up the challenge of this new way of working and thinking, and the ways of computing spread among us. . . . But times have changed. Apple has issued a new challenge—a challenge that explodes all of our comforting personal-computing standards and beckons us, once again, into the future.[116]

Rubin even addressed Apple's reframing of the Macintosh's black-box design head on. In order for the Macintosh to allow us "to forget about the physical computer," some "tradeoffs" were required "in terms of the 'standard equipment.' . . . Apple's challenge is for us to accept these differences in the interest of improved computing ability." Buyers therefore had to ask themselves whether they would "stick with the computers everyone else is using these days" or "accept the challenge and step boldly forward to where Apple asserts computing is heading."[117] While Rubin's review unhesitatingly approves of the Macintosh's design, even those reviewers who were critical of several aspects of its model of usability seemed tantalized by the promise of transparency. Anderson, for example, repeated the brochure's assertion that IBM's emphasis on improving the computer literacy of its users is the product of self-serving technocrats: "Certainly some users would prefer to be perceived as micro-Merlins. Perhaps the more cryptic a command code, the better. This category of user perceives the eventuality of real democratization of computer power with something akin to melancholia."[118] Williams, too, at times wrote as if he just walked out of an Apple shareholders meeting: "The Mac will delay IBM's domination of the personal computer market" by bringing us "one step closer to the ideal of computer as appliance."[119] Apple's rhetoric of usability thus exercised a significant influence over the reception of the Macintosh's software interface, leading many of its critics to ultimately reevaluate the machine's flaws through the conceptual framework the company had provided.

Conclusion

One repeated theme in this book is that usability is often approached as a way of resolving a conflict between computation and culture. The idea that there is such a conflict was a defining concern in some of the earliest articulations of personal computing as a model of usability separate from the scientific and corporate computing done on expensive mainframe and minicomputers. Even as hobbyists and some journalists assured us during the late 1970s and early 1980s that computers were moving out of the hands of the few into those of the many, that concern persisted. As I have shown in this chapter, journalists and corporate spokespersons began to reframe that conflict as a matter of computer literacy. Even though anyone willing and able to pay for one could now have access to their own, personal computer, industry observers suggested that most would have no idea what to do with it, let alone know how to control it. If the problem of computer literacy education were not addressed quickly and decisively, personal computers would become a source of stress and anxiety for most rather than tools of self-empowerment for all. As this chapter shows, many of the rhetorics of usability we deploy today to understand and engage with digital media are still implicitly framed in terms of how to avoid sliding back into a computer literacy crisis. Each new technology is somehow more transparent, more user friendly, or more convenient because the threat of computation's demands exceeding our ability to manage our lives is ever present.

Whereas IBM and Apple appeared to offer dramatically different solutions to the computer literacy crisis, today's personal computers offer a more uniform approach. There are some differences, certainly. That people continue to identify themselves as a "Mac person" or a "Windows person" shows how much influence the rhetorics and models of specific consumer technologies have on our computer literacy. People feel lost when moving from one platform to another, having internalized behaviors that feel natural on their system of choice but seem purposefully difficult or overly complicated when performed on the other. Given that Big Tech now designs not just personal computer operating systems but also enterprise infrastructure, cross-platform productivity software, phones, and media systems, it increasingly has the ability to shape our computer literacy not just with respect to personal computers but to almost every aspect of digital culture. Lev Manovich's discussion of how media-authoring software encourages users to adopt a mental model of art production based on the tools, scripts, and lexicon provided by the software's interface is helpful here.[120] The designers working for Apple, Google, Microsoft, and other companies developing widely used personal computing software may portray their work as transparent to human intention, but their technologies ultimately reshape the sociocultural aspects of computing around themselves. One way for us to consider alternative possibilities for design would be to pause and run our fingers over those seams that we do encounter despite the ubiquity of transparent design, tracing them to see where they lead and recognize the moments of decision they represent, rather than just slipping into the seamless experiences their rhetorics of usability encourage us to accept. Fortunately, there are growing efforts in science and technology studies to document these seams and even some consideration within the field of human-computer interaction of the value of user engagement with them.[121] Yet the continued emphasis on transparency in consumer software design suggests that the politics advanced by these researchers have not yet proven valuable to Silicon Valley.

As my case studies suggest, one reason transparent design has proven so durable in American personal computing is that technology journalism continues to contribute to a rhetoric of usability that positions transparency as the only conceivable approach to user-friendliness. Many high-profile technological publications continue to imply that there is no good way to address the needs of users that does not involve hiding the overly technical aspects of computation. In the increasingly rare instances when a new application does invite users to engage with complexity, such designs are coded in popular discourse at best as nuanced in ways that only a select group of power users can appreciate and at worst as unnecessarily oppressive. In many cases both past and present, technology journalists are content to interpret a given technology's model of usability using the rhetoric provided to them by its designers. Consider how in reviews skeptical journalists moved ultimately to accept the reasoning provided by designers in the case of the Macintosh or even invent their own justifications that were then taken up and amplified by the designer in the case of the PC. They approached these machines with concerns about the future of computing in mind, ones they internalized through a broader discourse unfolding across the publications they read and contributed to and that warned of a computer literacy crisis, but the PC and Macintosh were presented to them as solutions to that crisis, and they were happy to interpret them as such. Now consider how similar scenarios have likely unfolded—and are continuing to unfold—with respect to almost every piece of hardware and software produced since.

Reviews and breathless previews of computers are not the only type of pieces that technology journalists produce. In many of the sources I have cited here, critical discussion does occur. However, a substantial portion of the critical pieces found in technology journalism appears within a framework that accepts the cultural and social messaging of computing companies at face value. Thus, critical pieces often spend more time speculating over the future of the industry rather than reflecting on the political dynamics the industry imposes on us. So many articles both historically and today focus on ease of use and convenience or are written to fit new technologies within a recognizable narrative of innovation. There is a significant segment of technology journalism that accepts that narrative uncritically and teaches us to fit new rhetorics and models of usability into it rather than step back and challenge it. Just as transparent design keeps our focus on the task in front of us, many of the rhetorics and models of usability available to us today inculcate a form of computer literacy encouraging us to recognize the ease and convenience of computers and not consider other dimensions or consequences of computing. The academic research I have cited throughout this book is at least tacitly aware of the discursive power that designers wield, if not actively interrogating it. Several of these scholars have also begun to introduce their critical perspectives into journalistic contexts, and some technology journalists have started to engage with them.

I believe that a sustained, public discourse among academics and journalists about the cultural and social implications of technology is critical to challenging the self-serving rhetorics of usability that Big Tech works to maintain.

While the Macintosh is viewed as a foundational technology, drawing together many of the concepts I have traced in previous chapters into a rhetoric and model of usability that firmly associated transparent design with user-friendliness, it is not itself alone responsible for the durably of transparency. In the next chapter, I examine how many of the ideas that I have discussed up to this point were codified into a body of theory and methodology that is supported by scientific rather than merely market principles. That a transparent approach to user-friendliness could be justified in the long term by continued commercial success is to be expected, but its codification into a set of principles tied to accepted truths about the structure and limits of human reasoning has helped to ensure that its general goals have not been revised significantly over the past four decades even as the material conditions of personal computing have changed dramatically. There is a parallel intellectual history to transparent design's emergence in commercial hardware and software design that develops across fields like artificial intelligence, cognitive psychology, and computer science. However, as I show, many of the theorists who established the formal principles of human-computer interaction were strongly influenced by the developments they saw in the American personal computer industry. In this sense, their work served not just to explain these developments but also to elevate the focus on the consumer into a broad commitment to humanizing technology.

Chapter 5

The Human Factor

While in previous chapters I have examined the association between user-friendliness and transparent design in the American personal computer industry, in this one I consider how academic researchers came to view transparency as the foundation for an interdisciplinary approach to usability. Theory and research in the field of human-computer interaction (HCI) and the fields that have influenced its disciplinary self-definition are important sources for understanding the role that user-friendliness continues to play in American digital culture. Many of HCI's foundational concepts continue to inform the education of software developers, the professional ethics of user-experience experts, and the norms of practice in several areas of usability research. Foundational concepts in HCI also influence the lens through which many technology professionals consider the sociocultural aspects of computing. Academic research on usability in computing predates personal computing; however, as Martin Campbell-Kelly's history of the software industry notes, university curricula largely ignored personal computing in the United States until the mid-1980s because computer science departments focused primarily on preparing students to build and maintain software for corporate computing networks and office automation systems.[1] But while the consensus regarding transparent design in HCI research lagged a bit behind that forming among journalists and commercial software developers documented in the previous chapters, it would be inaccurate to say that HCI simply elaborated on their ideas. As I show in this chapter, HCI's articulation of transparent design has an intellectual and rhetorical history distinct from the popular discourse on user-friendliness. At the same time, there are a number of key resonances in the writings of early

HCI researchers that facilitated the significant influence that their ideas had on professional designers. HCI literature has helped provide a rhetoric of usability that leverages scientific principles to naturalize transparent design as the only way to build software that is culturally informed and socially engaged.

One could mark the beginnings of HCI with the founding of the Association for Computing Machinery's Special Interest Group on Computer-Human Interaction in 1982. However, as Elizabeth R. Petrick has shown, at least two different "parallel sets of citations" have been used to describe the field's history: one that focuses on the development of key technologies and another that tries to understand developments in computing in relation to other forms of media.[2] This chapter is positioned between them in that it traces how those researchers developing principles of usability understood the cultural significance and broader stakes of their work. To understand the intellectual heritage of the field, I use as my starting point the bibliographic histories prepared by HCI practitioners like John M. Carroll, Jonathan Grudin, and Brian Shackel. Each traces the origins of their field back to the late 1950s.[3] The histories they offer differ significantly from those found in digital media studies. Because humanists have focused on theories of usability that explicitly define computers as "media," they tend to position Vannevar Bush, J. C. R. Licklider, Douglas Engelbart, and Alan Kay as primary figures; their work is noticeably absent from Carroll's bibliography. And while these names do appear in Grudin's and Shackel's histories, they are identified only as thinkers who provided early "visions."[4] These three accounts describe HCI as an outgrowth of and response to "human factors" research in industrial engineering and as a new discipline produced through critiques of artificial intelligence and cognitive psychology that promised to offer a more interdisciplinary and socioculturally engaged approach to design. Despite these differences in perspective, many of the same questions about what roles computers should play in society that have animated discussions of usability in digital media scholarship also inform the rhetoric of usability developed in early HCI literature. In this chapter, I explore how foundational works in HCI theory defined their field as distinct from older human factors approaches to usability. Research by Ben Shneiderman, Donald Norman, Terry Winograd and Fernando Flores, Brenda Laurel, and Lucy Suchman promised to expand our understanding of what counted as use by enlarging the scope of human-machine relationships and recognizing the increasingly diverse purposes and contexts of use of personal computing.

Although each of these writers claims to foreground the diversity of identities, backgrounds, and interests that users bring to personal computing, I also argue that several also consistently appeal to naturalness, intuitiveness, or other essentialist qualities of human experience that effectively erase that diversity. Richard Coyne describes the trajectory of HCI's intellectual history as an effort undertaken by a rising cohort of technologists who sought to contrast their work with earlier "rationalist" approaches, promising instead to offer a comparatively "optimistic, enthusiastic, [and] utopian" vision of computing that "closely linked [their work] with the tenets of political and social liberalism."[5] However, Geoff Cooper and John Bowers' rhetorical analysis of early debates in HCI literature illustrates how the language early researchers developed to argue for expanding the scope of design considerations often left many key norms unchallenged, especially the idea that technical expertise enables professional technologists to describe how to realize users' needs better than users themselves.[6] While it is refreshing to see these engineers explicitly acknowledge the important role that critical and social theory can play in helping us to understand what is at stake in design beyond ease of use, it is important that we follow up and pay close attention to the actual principles of usability they propose alongside their calls to engage with pressing cultural, social, and political concerns associated with computing. In this chapter, I document a pattern of rhetorical moves through which Shneiderman, Norman, and Winograd and Flores treat users' identities and their contexts of use as high-level, abstract concerns and propose design principles that they claim are universally applicable because they address the low-level, psychological structures that define those concerns. From this perspective, heterogeneity may characterize users at an abstract level, but at a cognitive level all users share the same cognitive structures and constraints. However, this series of moves ultimately erases the diversity among users they claim designers must learn to address. Thus, while the rhetorics of usability they devise may draw on concepts from the humanities to emphasize the importance of cultural and social engagement during design, their frameworks in practice preserve older, normative practices.

In relying on frameworks that preserve normative practices, foundational literature in HCI recalls the ethics of expediency and clever hacks discussed in previous chapters. In their writings, HCI theorists similarly redefine the scope and scale of the contexts of personal computing, presenting their complexity as problems to be solved and narrowing their definitions of those contexts so that the comparatively simple, technologically driven solutions they propose will appear intuitively necessary. Throughout this book, I have used the terms "rhetoric of usability" and "model of usability" to refer to the ways that designers explain the significance of their design strategies and the material manifestations of those strategies, respectively. Because the writers I examine in this chapter are not creating technologies but describing theoretical frameworks to inform design practices, I refer to their schema as "principles of usability." As was the case with my discussion of how the politics of user-friendliness in commercial design emerged through the interplay of various rhetorics and models of usability, my goal here is to examine the interplay between how HCI theorists understand the sociocultural aspects of their work and the particular strategies that they claim will allow designers to be responsive to those understandings.[7]

These early HCI theorists, like the journalists, writers, and spokespeople whose positions I've outlined in other chapters, similarly regard problems of usability as resulting from a tension between computation and culture. Rather than address the complexities of this conflict, these theorists instead reframe the increasingly heterogeneous contexts of personal computing in ways that allow them to reclaim a degree of the homogeneity that had been assumed by the prior research they position their work against. While they pursue different lines of inquiry, I document how most invariably conclude that the best way to make computers seem universally usable and useful is not to develop a diversity of approaches to design nor to incorporate a diversity of perspectives into the design process but rather to design interfaces that are understood to complement inherent psychological principles, shared uniformly by all users. In advancing these rhetorics and principles of usability, the writers I examine in this chapter aim to connect their work with popular discussions about user-friendliness in the American personal computer industry. While these writers take a different path, their writings ultimately reach the same destination: transparent design becomes the natural solution to the sociocultural complexities introduced to usability by the rise of personal computing.

I begin this chapter by examining how the earliest approaches to studying the usability of computer systems emerged out of ergonomics, eventually drawing on research in cognitive psychology and artificial intelligence that describes a theory of "human information processing." These human factors approaches to usability were concerned primarily with improving the efficiency of workers who were understood to be specialist users, having received training on the specific tasks they would be responsible for when operating a computer as part of their job. I then examine how early theorizations of HCI show a repeated pattern of critiquing human factors approaches like the goals, operators, methods, and selections framework (GOMS) in order to position their work as more culturally and socially engaged. Ultimately, however, Shneiderman, Norman, and Winograd and Flores each embrace principles of usability that functionally reproduce the narrow focus of earlier research in usability that they view as problematic. While Winograd and Flores take additional steps to contrast their work with human factors computing research by developing a theoretical framework based on the thought of philosopher Hubert Dreyfus, their bricolage of concepts drawn from Western philosophy essentially reconstructs many key principles from human information processing theory and leads to conclusions very similar to those found in Shneiderman's and Norman's writings.

Following my review of these foundational texts, I then examine the work of two "outsiders" whose writings are cited by HCI theorists as evidence of the field's interdisciplinarity. In the context of this chapter, Brenda Laurel's and Lucy Suchman's writings are important tests, early moments when humanists and social scientists engaged with and responded to the rhetorics and principles of usability that HCI researchers claimed enabled their field to better address the sociocultural aspects of computing. Through comparing them, I consider the ways that concepts from the humanities can both be instrumentalized in support of the problematic politics of transparent design as well as play a key role in articulating a resistance to those politics. On the one hand, Laurel's work uncritically accepts many of the problematic principles I identify in the writings of Shneiderman, Norman, and Winograd and Flores. Drawing on the history and theory of Western drama, Laurel produces a rhetoric of usability that not only naturalizes transparency but represents designers as always exercising their authority in the best interest of users. Suchman's work, on the other hand, serves as a model for what a culturally engaged and socially responsible approach to design might look like. Rather than treat cultural and social concerns about computing as problems to be solved, Suchman suggests that the complexity and uncertainty they highlight should be embraced. Instead of claiming to provide an all-encompassing methodology for design, she describes design as an asymmetrical dialog that takes place asynchronously between designers and users through the interface. This imbalanced relationship not only establishes a boundary between the two but also affords designers a significant degree of control over that dialog. Whereas the other theorists discussed in this chapter seek to "screen out" user concerns by designing interfaces that engage with them on a "low level," Suchman engages with them more directly and thereby foregrounds the political dimensions of usability. In my conclusion, I briefly discuss the implications of Laurel's and Suchman's writings in light of more recent research on usability that is presented as attentive to the sociocultural aspects of computing.

Early Approaches to Usability

The first studies of the usability of computer systems focused on large machines used in industrial settings. The earliest research in human factors computing was published by Brian Shackel, who authored two ergonomic studies of large analog and digital computers in 1959 and 1962, respectively.[8] Shackel's research was primarily concerned with eliminating user errors by observing operators during use and identifying those components of interfaces where mistakes most frequently occurred. According to Grudin, early studies like Shackel's were influenced by the work of Lillian Gilbreth, whose holistic approach to time and motion studies is considered to be the foundation of modern human factors research in engineering, and Frederick Winslow Taylor.[9] Henry Dreyfuss was also an important figure in industrial design at this time, and his writings called for designers to shape interfaces around the motions, postures, and habits commonly adopted by people within specific use contexts.[10] Shackel's studies attempted to evaluate both the physiological and psychological strain of an interface's design, particularly the potential for user error based on misidentification of buttons, dials, and gauges when operating large control panels in laboratory settings. Most of the early work on computer usability thus pertained to "nondiscretionary" use.[11] Importantly, in nondiscretionary contexts all or most users share a basic level of training and experience, and the use environments are often highly structured toward specific purposes, which means that a considerable degree of homogeneity among users could be assumed.

A second wave of research on usability in computing emerged in the mid-to-late 1960s grounded in theories of human information processing drawn from cognitive psychology and artificial intelligence. Researchers in these two fields assumed that the human mind was on a fundamental level homologous to a digital computer and developed software decision-making systems to test hypotheses about human cognitive structures.[12] While there is a wide body of work describing human information processing models in both cognitive psychology and artificial intelligence, most writings cited in human factors computing research share a handful of features. Generally speaking, human information processing theory defines the human mind as a short-term processing module that serves as an environment for modeling problems, tasks, or situations. This short-term modeling module is connected at one end to a long-term memory module and to input / output modules on the other. This structure resembles the way that random access memory in computers serves as a temporary storage for data used during a computation process and the way that read-only memory—and later, physical storage media—serves to store the software that is drawn on to manipulate that data. Human cognitive behaviors are determined by sets of rules organized into hierarchies, with more common or generalizable rules mediating access to more specialized or narrowly defined ones. Additionally, most published research in this field assumed that human behavior is goal directed and that all human action is proceeded by the internalization of one's environment into one's short-term memory where it can be analyzed. The results of these analyses are then leveraged to develop goal-directed plans for action that draw on both those rules common to all human minds and those stored ones derived from an individuals' past analyses.[13] Many researchers further asserted that by understanding how these rule-based planning and decision systems functioned they could identify and develop strategies for working around inherent limits in human memory and attention. These insights could, in turn, inform the design and implementation of management systems that would be able to quantitatively measure the efficiency of individual and social behaviors in industrial and corporate settings.

The GOMS framework, developed over the course of the late 1970s and early 1980s by Stuart K. Card, Thomas P. Moran, and Allen Newell, is perhaps the most influential application of human information processing theory to the study of computer usability. Briefly, GOMS is a holistic framework for designing software systems that complement human cognitive structures. It provides a framework for designers to simulate user decision making in order to make judgments about and evaluate the efficiency of system design. As the name implies, the GOMS framework is based on human information processing theory's broad assumption that all human behavior is goal directed and proceeds according to plans that human beings operationalize by selecting from and sequencing available methods. In *The Psychology of Human-Computer Interaction*, Card, Moran, and Newell are quite explicit in their reliance on human information processing theory, stating directly that their work is founded on the belief that "certain central aspects of computers are as much a function of the nature of human beings as of the nature of the computers themselves."[14] They understand that programming languages and interfaces are designed at the intersection of computer science and psychology. Thus, their vision for GOMS is to establish a framework for design that articulates psychological principles "in terms homogenous with those commonly used in other parts of computer science." They explain that such a framework is necessary for computer engineers because it is unlikely that psychologists will become "primary professionals in the field" of engineering. Nor should computer scientists be expected to retrain themselves as psychologists.[15] Significant portions of the book are thus dedicated to a method for describing user behavior through a pseudocode that they claim allows designers to simulate how humans cognitively deconstruct their abstract goals into operators, methods, and selections. This pseudocode can also be used as a sort of blueprint for software developers to follow when they are designing the structure of the menus and command sequences they build so that the menus and command sequences will complement the cognitive behaviors that they mapped out during their simulations.

GOMS became popular and continues to be applied within some areas of usability research because it provides a framework for systemically quantifying human behavior via an integrative model for user observation. Because it assumes that all users will possess a universally shared set of cognitive structures and behaviors, GOMS is able to reframe user errors as the result of inefficient interface and system design. In order to help developers systematically evaluate the usability of the software they build, Card,

Moran, and Newell provide mathematized descriptions of human perception, motor skills, simple decision making, learning and retrieval, and problem-solving strategies. The cognitive and physical actions that users undertake can be enumerated by collecting data like the number of commands they issue to a computer system, by noting the time it takes them to perform a task, and by tracking the number of keystrokes they make. Following quantification, users' data can be integrated into a singular model of human information processing, allowing software developers to evaluate the efficiency of their designs by identifying and explaining trends visible across the collective data of all users.[16] Yet even as they promise that their method will ultimately improve the efficiency of a technology for all users, Card, Moran, and Newell do admit to limits to the universal applicability of their model. When interpreting the results of their demonstration case studies, they explain that they observed a difference between expert and nonexpert users of "about a factor of three." Additionally, something as seemingly simple as variations in "the user's typing speed" will necessarily have an effect across all measurements.[17] Implied in their conclusion to *The Psychology of Human-Computer Interaction*, in short, is that GOMS becomes less effective the more widely varied a system's users are. As I discuss, this limitation is one that other, rising researchers in usability would attribute to an overreliance on human information processing theory and point to as they called for a new approach to design that could account for the increasingly diverse contexts of personal computing.

A New Approach to Usability

At the same that GOMS was being developed, several researchers began to recognize that those principles of usability that assumed significant homogeneity among users disincentivized the consideration of anything beyond the immediate cognitive and bodily relationship between user and machine. Models of nondiscretionary usability were accompanied by hiring and training procedures that normalized user behavior so that most differences among users could be in practice ignored. Critics of human information process theory like Ben Shneiderman and Donald Norman argued that designing software for personal computers required a new approach to usability because the contexts these machines would be used within could not be as tightly managed. In short, designers needed to find a way to account for the wide range of identities, interests, knowledges, skills, and purposes that discretionary users would bring to personal computing. As I show in this section, however, the commitment to engage with the complex sociocultural aspects of personal computing that they highlight in their rhetorics of usability is not reflected in the principles of usability they define. In practice, their work ignores the very concerns they raise when critiquing older approaches to design and in doing so serves to rehabilitate essentialist understandings of learning and memory in ways that "screen out" diversity among users.

Before he began raising concerns about the limits of human information processing theory, Shneiderman had himself been writing about its application to computer science, specifically in the design of programming languages and in programming pedagogy. For example, his 1980 textbook, *Software Psychology*, reviews much of the then current research that applied theories and methods from cognitive psychology to software design, even citing several of Card, Moran, and Newell's early articulations of GOMS. In the book's final chapter, however, Shneiderman calls on students to pay attention to the social implications of computing by drawing on several critics of American technoculture, including Lewis Mumford, Theodore Roszak, and Joseph Weizenbaum. He explains that he sees a "gaping chasm between social commentators who perceive technology, particularly computer technology, as dreadfully harmful and computer science researchers who feel that computer technology can lead to a better way of life."[18] Shneiderman suggests that neither side in this debate is necessarily correct. Both sides are animated by the perception that humans and computers are "competing for the same ecological niche"; however, as time goes by, "the dichotomy between human creative skills and the computer's tool-like nature will become more clear."[19] If we expand the scope of design, he suggests, we can build computers and develop software that will prevent computational practices from competing with or threatening cultural ones. Resolving this conflict would thus require designers to, among other things, "consider the impact of systems on people's lives" and "treat people as individuals."[20] This conflict, he concludes, is one that would define usability for years to come.

Shneiderman would go on to propose principles of usability intended to address this tension between computation and culture with his theory of direct manipulation interfaces. When first introducing these principles, Shneiderman explains that this new approach to design is one that foregrounds "the expansion of the user population to include novice and non-technically-trained people," groups whose needs are "so vastly different that the experience and intuition of senior programmers may be inappropriate."[21] Rather than try to wholly anticipate user behaviors as frameworks like GOMS did, Shneiderman instead calls for designers to create interfaces that would allow users to learn through nonlinear experimentation instead of rote memorization of commands. On a very basic level, Shneiderman describes the advantages of a direct manipulation system as follows:

1) Novices can learn basic functionality quickly, usually through a demonstration by a more experienced user.
2) Experts can work extremely rapidly to carry out a wide range of tasks, even defining new functions and features.
3) Knowledgeable intermittent users can retain operational concepts.
4) Error messages are rarely needed.
5) Users can immediately see if their actions are furthering their goals and if not, they can simply change the direction of their activity.
6) Users have reduced anxiety because the system is comprehensible and because actions are so easily reversible.[22]

Unlike GOMS, the goal of direct manipulation is not to boost the efficiency of interaction. Instead, Shneiderman explains, designers need to build a tolerance for mistakes into interfaces so that users will feel more comfortable working through their imperfect understandings of a system. In other words, users should be able to learn how to use an application through low-stakes experimentation. This approach does initially seem to support increased user agency in the sense that it affords users more control over how they learn new use behaviors. However, if we look more closely at Shneiderman's principles of usability, it quickly becomes apparent that they are grounded in several problematic assumptions drawn from human information processing.

While direct manipulation interfaces marked a break from the then current norms of usability theory, Shneiderman argues that were there already numerous examples of his principles succeeding in other familiar technologies. The auto industry, for example, had already effectively addressed many of the same problems that personal computer was facing as evidenced by how commonplace driving had become in American culture:

> The scene is directly visible through the front window and actions such as braking or steering have become common knowledge in our culture. To turn to the left, simply rotate the steering wheel to the left. The response is immediate and the scene changes, providing feedback to refine the turn. Imagine trying to turn by issuing a command "LEFT 30 DEGREES" and then having to see where you are now—but this is the level of many [personal computing] tools of today.[23]

We learn how to perform fundamental driving maneuvers by doing them, practicing until we can accomplish a complex series of actions without having to stop and consider each movement. After a brief period, driving seems natural to most drivers. Additionally, advances in automotive engineering have further reduced the physical and cognitive demands of even the simplest of maneuvers. Within computing, Shneiderman identifies several similar examples of applications that he believes already support a more intuitive form of learning like VisiCalc, text editors, and computer-aided design and drafting applications. Video games, he notes, appear to offer the clearest example of direct manipulation in the way that they "provide a field of action which is simple to understand since it is an abstraction of reality—learning is by analogy. Watching a knowledgeable player for several minutes is sufficient to learn the basic principles, but there is ample complexity to entice coins from experts for many hours."[24] Direct manipulation, in short, has the potential to accommodate a wide variety of user interests, needs, and skill levels. Even users with an advanced degree of computer literacy, Shneiderman suggests, could potentially benefit, as new methods for providing increased user feedback could also improve their ability to execute complex tasks.

If we look more closely at Shneiderman's descriptions of how user behaviors are learned through direct manipulation, it soon becomes apparent that his understanding of how learning would occur through this new style of interface is based on concepts from his earlier research applying human information processing to programming pedagogy. As he explains, direct manipulation interfaces are able to support a wider variety of users because they offer semantic representations of computation: abstract "high-level concepts in the problem domain" that are "largely system independent" and that users can learn to decompose "in a top-down way into multiple lower-level concepts." Semantic knowledge provides a general understanding of a given problem space, which serves as a framework for action and is transferrable to new instances of similar tasks. Frameworks like GOMS, on the other hand, prioritize syntactic knowledge, understandings that are "system dependent," largely arbitrary, and can only be acquired through "rote memorization."[25] Whereas a semantic representation of how to write a certain kind of text would include norms of document design and genre, broken down to specific features of style and sentence structure, a syntactic representation would include the specific commands needed to set indentation, make typographical adjustments, or save the file to disk. Shneiderman explains quite directly in an earlier publication that he coauthored with Richard Mayer that they originally defined these two concepts in the context of computing as part of attempt to develop a "unified cognitive model of the programmer" based on human information processing theory.[26]

In the course of describing the roles that semantic and syntactic knowledge play in several different programming-related tasks, Shneiderman and Mayer make two observations that are important for Shneiderman's later work on direct manipulation. First, they point out that programmers typically draw on semantic knowledge to understand problems and syntactic knowledge to implement solutions. As a consequence, new syntactic knowledge is often developed through the application of prior semantic knowledge.[27] Shneiderman applies this idea to usability when he describes how novices would be able to more quickly memorize commands and action sequences if they were presented through direct manipulation interfaces that appear to function as "appropriate physical model[s] of reality" rather than as computational processes. Working from their semantic understanding of a "'Rolodex'-like device," to use Shneiderman's example, users could apply a form of top-down reasoning to experiment with the system syntactically. Over time, they would identify command sequences that would allow them to flip through the cards displayed on screen in search of information in ways that would feel similar to their manipulation of a physical Rolodex.[28] Furthermore, even in cases in which there were no obvious physical parallel, Shneiderman argues that direct manipulation could still be realized if interfaces presented users with a recognizable causal logic that they could draw on to evaluate their actions immediately.

Second, Shneiderman and Mayer suggest that semantic and syntactic knowledge are largely separate and observe that it is easier for programmers to develop new syntactic understandings than it is for them to develop new semantic ones. In other words, it is easier to figure out how to perform a new task with an unfamiliar piece of technology than to understand how that technology works or why it is structured in a specific way. As constructs of long-term memory, they explain, semantic understandings are widely applicable because they are a form of integrative knowledge. Abstract, high-level semantic understandings often overlap, however, because they integrate low-level understandings into abstract, meaningful rules for interpretation and decision making. Their integrative nature makes them durable but also hinders the formation of new abstract understandings, as each new one must be reconciled with those that already exist in a user's mind.[29] While these two points may seem insignificant outside of academic debates about the mechanics of the human memory and attention, they have significant implications for the power dynamics between software developers and users.

Shneiderman acknowledges in his application of these observations to direct manipulation that users' leveraging of semantic understandings of noncomputational activities may interfere with their ability to develop new semantic understandings of computational processes. He admits that modeling interfaces after noncomputational processes "may be misleading. Users may rapidly grasp the analogical representation but then make incorrect conclusions about permissible operations" by assuming that the computer internally functions like the processes that it visually models. To avoid confusion, designers must "selectively screen out" references to computation so that users are better able to interpret their actions within the more familiar frameworks presented to them on screen.[30] In other words, users of direct manipulation systems would have trouble integrating their semantic understandings of the metaphors visible on screen with later attempts to develop semantic understandings of the computational systems that underly those metaphors. Most users might only be able to learn to think and talk about computers as if they really were just spreadsheets or documents or digital Rolodexes and not also as systems of algorithms and data structures. To avoid confusion among users, an effective direct manipulation interface should therefore intentionally minimize exposure to the specific computational operations that users execute through them. Appealing to what he assumes are inherent limits to human learning, Shneiderman thus suggests that the advanced degree of computational literacy that he exercises is not one that well-designed interfaces should encourage among users.

Users, his writings imply, have more important things to reserve their cognitive labors for.

A similar reframing of human information processing as part of a call for software developers to consider the broader sociocultural aspects of design can be found in the work of Donald Norman. While Shneiderman came to questions of usability from computer science, Norman comes from cognitive psychology. Prior to his research on usability, Norman's research and writings focused on human memory and attention.[31] Initially, his motivations for studying usability were to strengthen what he felt were critical limitations in research on human information processing in psychology.

For example, in a 1980 essay titled "Twelve Issues for Cognitive Science," Norman argues that existing theories of human information processing were not able to account for the "social and culture factors" and other "major points that distinguish an animate cognitive system from an artificial one." Most treat the human mind "as pure intelligence, communicating with [other minds] in logical dialogue, perceiving, remembering, thinking where appropriate, reasoning its way through the well-formed problems that are encountered in the day. Alas, that description does not fit actual behavior."[32] Although humans share basic cognitive structures, they do not share the same experiences, memories, and embodied identities. Norman asks his readers to consider the context of a classroom, a familiar environment that quickly becomes grossly complex if we step back from the ideal cognitive model of a teacher speaking to student-listeners who process the spoken information. The students and the teacher both simultaneously participate in and perceive the classroom environment, the events that unfold within it, and their relationships to one another, but each of them experiences it differently as evidenced by the inequality of student outcomes. Norman thus concludes the essay by calling on cognitive scientists to avoid "narrowly-based research . . . within a vacuum" and instead turn toward "goal directed, conceptually based research" that will plunge them "as deeply as possible into the tangled web of specific problems that exist within any area of concentration. Then, let the results drive the investigation, so that the studies become the driving force for further research."[33] Although Norman is critical of human information processing theory here, he is also quite explicit in his suggestion that it could be revised to account for the complexities of contextualized cognition. In the years since, Norman has reinvented his professional profile as a usability expert; however, it is important to recognize that this pivot to usability was motivated by his goal of finding new research contexts for developing a more elaborate understanding of human information processing.

Building on these criticisms of research in cognitive psychology, Norman developed a rhetoric of usability that situates design as a discipline that mediates between cognitive and cultural representations of activity. For Noman, usability is a field dedicated to understanding how humans internalize models of action "in the head" by interpreting the affordances and constraints presented to them "in the world" by an interface. In contrast to Card, Moran, and Newell, Norman argues that the goal of usability should be "systems that are pleasant to use—the goal is neither efficiency nor ease nor power, although these are all to be desired, but rather systems that are pleasant, even fun."[34] Norman leaves "pleasant" undefined here, but his examples suggest that unpleasant experiences are those where users struggle to internalize the mental model that designers had intended their technologies to support. Even the simplest of mechanisms, he explains, can seem confusing and lead to a sense of discomfort in the operator if "there is a discrepancy between the person's *psychologically* expressed goals and the *physical* controls and variables of the task." These discrepancies form the basis of a pair of usability principles that Norman applies throughout much of this later work. The first is the "gulf of execution," which represents the cognitive effort required to undertake an action sequence in pursuit of a desired goal. The second is the "gulf of evaluation," which represents the cognitive effort needed to interpret the results of that action sequence.[35] Creating usable interfaces demands that one or both gulfs are bridged

> by starting in either direction. The designer can bridge the Gulfs by starting at the system side and moving closer to the person by constructing input and output characteristics of the interface so as to make better matches to the psychological needs of the user. The user can bridge the Gulfs by creating plans, action sequences, and interpretations that move the normal descriptions of the goals and intentions closer to the description required by the physical system.[36]

Usability, in short, is a field dedicated to aligning a user's interpretation of a model of usability with the designer's intended interpretation. As Norman notes elsewhere, this alignment can occur in a variety of ways—via formal instruction or training, through conversations with other people about the system, or by referring to documentation. For Norman, however, the interface is the only source that can be counted on to significantly influence the majority of users.[37]

Norman explains that when designers are developing software, they can more easily manage these two gulfs if they design interfaces that prioritize the psychology of the user over accurate representation of the entirety of a computer system. Yet, he explains, influencing how users construct mental models is a very complicated problem: "Not only do users differ in their knowledges, skills, and needs, but for even a single user the requirements for one stage of an activity can conflict with the requirements for another."[38] Complicating matters further is the fact that "the number of variables and potential actions" influencing how a user might interpret an interface "is large, possibly in the thousands."[39] Designers, he argues, have previously approached the problem of user comprehension as one of providing enough information or of properly organizing information, but "extra information can impair intention selection, in part by providing distractions." However, not providing enough information can also leave users bewildered.[40] Good interfaces, in other words, are ones that understand "which messages [about the system] to give" to the user and "which to defer" at any given moment. A good interface should also "know where on the screen messages should go without interfering with the main task" and recognize when "associated information . . . should be provided." Citing what he believes to be inherent, universal limits to human memory, Norman argues that anything that appears on the screen is a potential distraction and that any new information must be "presented at the right time, at the right level of specification."[41] Ultimately, the simplest way for designers to influence users' interpretations of models of usability is to carefully control the amount and kind of information provided to them.

Norman's later work defines principles of usability that position simplification as the ultimate goal of design. Unlike mechanical devices, interfaces for computer systems have very few built-in affordances or constraints, apart from the standards that influence the layout of a keyboard or the number of buttons on a mouse. The relationship between screen and algorithm is arbitrary: software functions "electronically, invisibly, with no sign of the actions it is performing" unless specifically instructed by a programmer to make them visible.[42] This softness may seem challenging, as it means that designers are responsible for determining the entirely of a computer system's representation. However, this challenge also provides designers with a unique opportunity: "The computer can be like a chameleon, changing shape and outward appearance to match the situation. The operations of the computer can be soft, being done in appearance rather than substance," allowing software to "take on the appearance of the task" so that a system's technical complexities can "disappear behind a façade."[43] In other words, Norman calls on designers to radically simplify the experience of using a computer by strategically misrepresenting its internal functions so that software more readily directs users toward designers' intended mental model of a particular task. Norman stresses the need for design to normalize behavior much more than Shneiderman does, but he similarly concludes that the best approach to design is to develop an alternative representation of computing that supports a simpler, less technically oriented form of computer literacy.

Notably, both Shneiderman and Norman define principles of usability that position a narrowed form of computational literacy as a natural solution to inherent constraints on human cognition. Their solutions are in this respect not unlike those proposed by journalists during the early 1980s and realized by the Apple Macintosh, as described in the previous chapter. Unlike rhetorics of usability documented in previous chapters, however, this approach is not intended to make computing more appealing to skeptical consumers who do not have time for or the interest in learning about how computers work. Rather, Shneiderman and Norman appeal to scientific justifications, suggesting that good design must account for and develop interfaces that complement what they believe are inherent limits to users' ability to achieve a balance between the cognitive demands of computation and those of the cultural and social activities they are using a computer to accomplish. On the surface, Shneiderman and Norman seem to call on designers to find ways to be responsive to concerns about user identity and social behavior resembling those that shape inquiry in the humanities. Yet neither theorist actually engages with the concerns they raise in their rhetorics of usability. The conflict between computation and culture they cite as a problem for software designers can, they ultimately suggest, be solved more simply by reframing the high-level, abstract concerns regarding users' identities and their contexts of use in terms of the low-level, inherent cognitive structures that all users share, regardless of any differences among them. Thus, both Shneiderman and Norman reimpose homogeneity onto users, rehabilitating the very frameworks they criticize as inattentive to the diverse needs of users.

The "Phenomenological" Turn

A generous reading of Shneiderman and Norman may conclude that the tensions between their rhetorics of usability and the problematic assumptions in their principles of usability are due to a hasty appropriation of language from other fields, leveraged to provide their work a sense of exigence when compared to other approaches to usability. But as I show in this section, other early foundational HCI texts reached similar conclusions despite engaging more directly with theoretical frameworks from the humanities. In 1965, at the request of the RAND Corporation, a philosophy professor named Hubert Dreyfus issued a report examining the key problems facing researchers in artificial intelligence.[44] Dreyfus would later publish an expanded version of the report in 1972 as *What Computers Can't Do: A Critique of Artificial Reason.*[45] Although Dreyfus's writings were largely ignored among computer scientists when they were first published, Terry Winograd and Fernando Flores cite his work as a key influence on their 1986 book, *Understanding Computers and Cognition: A New Foundation for Design.* Dreyfus is also among those that Winograd and Flores thank in their preface for offering commentary on drafts of their book.[46] Given the influence of artificial intelligence on early research on usability via theories of human information processing, one could view Winograd and Flores's work as an update to and extension of Dreyfus's work. Yet as I show in this section, Winograd and Flores develop a rhetoric of usability modeled after Dreyfus's critique of artificial intelligence but propose principles of usability homologous to Shneiderman's and Norman's that similarly prioritize low-level understandings of human experience in order to simplify and resolve tensions between computation and culture at a high-level. In short, they piece together a framework that leverages concepts from Western philosophy, semiotics, and cybernetics with the goal of naturalizing transparent design and thereby facilitating the realization of universal usability.

While Dreyfus's writings are still worth reading in their entirety today, two of his criticisms are particularly important for understanding his influence on Winograd and Flores. First, he argues that research in artificial intelligence often explodes the significance of what are narrowly defined, technically oriented problems. He describes many of the studies he reviewed as full of logical leaps, with a "rhetorical presentation . . . often substitut[ing] for research, so that research papers resemble more a debater's brief than a scientific report."[47] Dreyfus does not dispute "the importance and value of their research on specific techniques [in computer science] such as list structures, and on more general problems such as database organization and access, compatibility theorems, and so forth"; however, he is skeptical of the human-computer analogy that researchers have relied on to explain the significance of their findings outside the field.[48] Dreyfus argues that artificial intelligence researchers have treated the analogy as axiomatic and suggests that they have had little incentive not to given the Western bias toward reason as the essence of human intelligence. Scientists since Galileo, he continues, have interpreted the history of philosophy dating back to Plato as a series of efforts to abolish "all appeal to intuition and judgment."[49] As a result, the field of artificial intelligence, he claims, has been largely uncritical of Alan Turing's assumption that computers would one day perform operations that would be indistinguishable from human intelligence to observers. They see their field as dedicated to realizing this analogy rather than determining whether it is even true in the first place.[50]

Additionally, Dreyfus argues that many of the problems that artificial intelligence has struggled to solve serve as evidence that there are aspects of human intelligence and experience that cannot be accounted for by models of detached reasoning. Dreyfus observes that a significant percentage of artificial intelligence studies conclude that any behaviors that cannot be accounted for by an otherwise successful cognitive rule set could be addressed by the development of additional rules. This continual adding of rules upon rules to account for new, previously unmodelable behaviors inevitably produces "a regress of rules for applying rules." At some point, "the lowest level rules are automatically implied without instructions," with the result that many basic assumptions in artificial intelligence are treated as unassailable truths because without them the newer rules animating current research would lose their validity.[51] Even were this not a concern, Dreyfus further argues that artificial intelligence's commitment to rule-based models of cognitive behavior does not account for the fact that humans often act on imperfect information. The formalized systems of rules that artificial intelligence researchers devise rely on the impossible ability of a human cogitator to attain an objective perspective on a situation wherein all relevant information is either available or discernible through rule-based reasoning. In practice, the significance of objects, intentions, and utterances are often ambiguous. Selecting a context for interpretation would thus produce an infinite regression of rule selection. Initially, a set of rules is needed so that users can select rules for interpretation, but ambiguities will inevitably arise, requiring the application of yet another set of rules to address the uncertainty, leading to still more instances of ambiguity and so on.[52]

While computer scientists have struggled to develop rules capable of handling the ambiguity of human experience, Dreyfus notes that humans often barely seem to notice it. Referencing Martin Heidegger's phenomenology, Dreyfus explains that humans are quite adept at managing this complexity because "we are always already in a context or situation which we carry over from the immediate past and update in terms of events that in the light of this past situation are seen as significant."[53] But computers, he argues, cannot operate in this way. The rule-based models of computational cognition in development during the mid- to late 1960s had to discretely analyze every possible detail in a situation, which meant that placing computers in even the simplest, most ordinary contexts of human experience could lead to a state of paralysis. Even if one takes into account the ways that human information processing theory tries to transform experience into rules stored in long-term memory, he asserts, it is still impossible to simulate the totality of experience that humans effortlessly and continuously apply to the ambiguities they encounter. The experiments of 1960s research in artificial intelligence, he concludes, only appear to validate the human information processing model because they are conducted within narrowly defined contexts that carefully control, minimize, or avoid the kinds of complexity encountered in everyday human experience.

As if picking up where Dreyfus left off, Winograd and Flores insist over a decade later that usability needs to be an interdisciplinary endeavor that resists privileging scientific approaches over sociocultural views of technology. Much like Dreyfus, they insist that frameworks relying on human information processing theory like GOMS are "based on a misinterpretation of the nature of human cognition and language. Computers designed on the basis of this misconception provide only impoverished possibilities for modeling and enlarging the scope of human understanding."[54] Theories of usability that are based on human information processing only consider the "functional" success of their methods, but to address the new variety of contexts that computers will soon be found in, usability must also "incorporate a holistic view of the network of technologies and activities in which [computational activity] fits, rather than treating the technological devices in isolation."[55] Like Dreyfus, they find the tendency among computer scientists to frame user behavior as a kind of detached reasoning unsurprising. The scientific tradition has taught most designers that objectivity, understood as rational thought that is removed from "subjective" concerns like identity, affect, and social norms, is considered to be "the very paradigm of what it means to think and be intelligent. . . . For someone trained in science and technology it may seem self-evident that this is the right (or even only) approach to serious thinking."[56] Yet, they observe, the emphasis on user-friendliness in 1980s commercial design and personal computing advertising is evidence that this mindset is not the way most people engage with computing.[57] Usability research and software design, they conclude, need to incorporate theoretical perspectives that can help to situate computers more firmly in frameworks that take into account how humans experience and interpret the world around them.

Following Dreyfus's lead, Winograd and Flores develop a rhetoric of usability that positions hermeneutics and phenomenology as a foundation for a more human-centered approach to design. Along with Dreyfus's commentaries on Heidegger, they also draw on Humberto Maturana's autopoiesis, C. S. Peirce's semiotics, and the speech act theories of J. L. Austin and John R. Seale. Leveraging this interdisciplinary framework, Winograd and Flores argue that their principles of usability are more representative of "what it means to exist as a human being, capable of thought and language." While each of these theorists plays a key role in their criticism of human information processing, Heidegger's concept of being-in-the-world serves as their basic framework for understanding human experience.[58] Winograd and Flores posit that "the practices in terms of which we render the world and our own lives intelligible cannot be made exhaustively explicit." We exist in a state of "thrownness" such that we are always moving forward in time. We are always already in a context or situation that is subject both to the past influences we carry forward into the present as well as the histories of the objects and people around us. As a consequence, they continue, we experience consciousness as being-in-the-world, meaning that we cannot act in the world and reflect on the nature of our actions at the same time. It is therefore impossible to enumerate every meaningful detail we experience or every reason behind a decision that we make. In practice, "existence is interpretation and interpretation is existence" in the sense that we act largely on "what people loosely call . . . 'instincts.'" Because it is impossible to fully explain or to explicitly reason through every detail of a decision while acting, neither our understanding of the world nor our sense of self is stable. Nonetheless, we often experience moments of "breakdown" when our being-in-the-world is interrupted. In these moments, we cease acting. The context we exist within then becomes "present-at-hand" and is explicitly knowable, at least partially, through reflection. While breakdowns are disruptive, the moments of reflection they create can be productive if they afford us a chance to develop alternative understandings of our actions, our sense of self, or our relationship to the world that we were not able to recognize while acting.[59] Within this framework, Winograd and Flores conclude, the kind of detached reasoning described in human information processing theory is only possible during moments of breakdown. Programmers engage in a form of detached reasoning when designing software—as they must enumerate and operationalize every aspect of what they model in source code—but they must recognize that breakdown is a state of exception within human experience, not the norm. The goal of usability, they suggest, is to find a way to engineer computer hardware and software to reproduce a sense of being-in-the-world, avoiding moments of breakdown that would push users to cease acting and require them to reflect on the technology before them.

Winograd and Flores close their book by defining principles of usability for an "ontological" approach to design that would allow users to be-in-the-world during use by experiencing a "transparency of interaction." They assert that the most successful systems—technological and social—are ones that are able to account for breakdown without interrupting the flow of human action. Breakdown is in some sense inevitable, especially in the context of human communication; however, software developers must approach design holistically as a project of "creat[ing] the world in which the user operates."[60] Winograd and Flores point to the automobile as an example of how designers can create a world where interpretation is smoothly integrated into experience:

> In driving a car, the control interaction is transparent. You do not think 'How far should I turn the steering wheel to go around that curve?' In fact, you are not even aware (unless something intrudes) of using a steering wheel. Phenomenologically, you are driving down the road, not operating controls. The long evolution of the design of automobiles has led to this readiness-at-hand. It is not achieved by having a car communicate like a person, but by providing the right coupling between the driver and action in the relevant domain.[61]

This example is striking because it so closely resembles Shneiderman's own illustration of direct manipulation, both through its reference to automobiles and through the way it positions transparency as a form of interaction that users experience as natural. Despite positioning their work as a distinct break from conventional usability theory, Winograd and Flores in moments like these demonstrate that their framework is effectively a recreation of earlier models, cobbled together from concepts drawn from the humanities.

Like Shneiderman's and Norman's, Winograd and Flores's rhetoric of usability places computation and culture in conflict with one another and suggests that the tension between them can only be resolved by principles of usability that restrict engagements with technical concepts to specialists. This perspective can be clearly seen in their example of electronic mail systems. They explicitly state that the user interface provided by the client software should reveal as little as possible about the mail servers they communicate with in order to avoid "intrusion[s] from another domain—one that is the province of system designers and engineers." The domain of the user, if it is to be "ontologically clean," should be "constituted of people and messages sent among them."[62] When encountering a breakdown, error messages should avoid reference to any fault or condition of the server. If a user's message were to fail to send, a response like "Cannot send message to that user. Please try again after five minutes" is thus preferable to "Mailbox server is reloading." The latter contains a technical term that is not a part of users' normal work experience and encountering it would prolong their state of breakdown rather than quickly return them to being-in-the-world. It is crucial to recognize here, however, that for Winograd and Flores phenomenological transparency as an approach to design is ultimately a way to reify existing social structures. Despite spending much of their book talking about designing software to complement natural experiences, their concluding examples focus almost exclusively on ways that computers could be designed to invisibly integrate managerial oversight and improve productivity in ways that would prevent workers from actively reflecting on the way software is restructuring their thinking and doing. Ontologically clean, transparent interfaces make it easier for knowledge workers to do their jobs by blocking them from interrogating the tools they use to carry out their duties.

Given that Winograd and Flores position their principles of usability as an extension of Dreyfus's critique of artificial intelligence, it is worth reading the two against one another. That their ontologically clean approach would have the functional benefit of increasing the ability of novice users to perform specific tasks is hard to challenge. Yet, irrespective of the practical benefits of the ontologically clean interfaces, Winograd and Flores make claims that seem to go beyond merely offering recommendations for how technology ought to be constructed. In adopting an ontological approach to design, they are, they admit, "asking more than what can be built. We are engaging in a philosophical discourse about the self—about what we can do and what we can be. Tools are fundamental to action, and through our actions we generate the world. The transformation we are concerned with is not a technical one, but a continuing evolution of how we understand our surroundings and ourselves—of how we continue becoming the beings that we are."[63] Certainly there is more truth to these claims than those "debater's briefs" that Dreyfus objected to in artificial intelligence research, especially when considered from our present moment when personal computers have been integrated into almost every facet of our lives. However, the restrictions on access to computational systems by users that Winograd and Flores advocate for is one that leaves designers in control of that transformation.

As if anticipating that criticism, Winograd and Flores insist that designers have only so much influence over the implementation of technology, giving users a place to determine the direction it takes in our society. Such an opening, however, would seem to contradict the very separation between computational and cultural domains that they insist is necessary for computers to become widely usable. In this respect, Winograd and Flores's principles of usability enact a politics of design functionally similar to those proposed by Shneiderman and Norman. Their engagement with humanistic principles is here instrumentalized in pursuit of a model of usability that gives designers the sole authority to determine the shape and structure of our relationship to technology. The goal of design is thus to subtly normalize our thinking and behavior so that the intended ways of knowing and doing supported by a technology can come to be experienced universally as natural.

Outsider Approaches to Design

HCI's promise of interdisciplinarity would also draw in researchers from outside computer science and cognitive psychology. Brenda Laurel's and Lucy Suchman's writings are both influential in different ways, placing theories and methods from the humanities and social sciences into an extended dialog with many of the ideas proposed by HCI theorists. Laurel's work in particular serves as a point of direct connection between many of the theorists I have discussed so far. For example, a collection that Laurel coedited with S. Joy Mountford, then manager of Apple's Human Interface Group, includes contributions from Grudin, Kay, Shneiderman, and Norman, as well as other notable observers and industry figures I have not mentioned, such as Nicholas Negroponte, Ted Nelson, Howard Rheingold, and John Walker. When referenced by engineers and cognitive psychologists, Laurel's writings are treated as proof of HCI's interdisciplinary mission.[64] Laurel's own writings are often viewed as one of the first, sustained engagements from the humanities with digital media and software design. Suchman's ethnomethodological approach to studying usability helped expand user observation in HCI through the adoption of more qualitative methods. An anthropologist by training, her early writings discuss observations she made of staff at Xerox's Plato Alto Research Center interacting with copying machines. As I argue in this final section, there is a distinct difference between how Laurel and Suchman apply these outsider frameworks to the study of usability. In her extended engagement with HCI through the lens of the history and theory of drama, Laurel ends up affirming many of the problematic assertions about the relationships between users and designers I have outlined in this chapter. Suchman, by comparison, shares some views of human behavior with early HCI theorists but resists the urge to offer simple solutions. Rather than treat usability as a problem to be solved, she encourages designers to see usability as a continuous exploration of the cultural and social relationships between users and designers.

Like those works discussed so far, Laurel's *Computers as Theatre* begins by arguing that current research in usability has to date been too focused on narrow definitions of human experience.[65] While Laurel doesn't refer to Dreyfus or Winograd and Flores, she does similarly position her work as a response to research that relies on models of detached reasoning. If we are to realize the "intrinsically interdisciplinary nature of design," she begins, then we need to "blur the edges between application and interface" by similarly blurring the edges between computation and culture.[66] Although she notes that by the end of the 1980s, some HCI researchers had begun to incorporate concepts from "artistic disciplines . . . as elements of good design," in practice HCI literature only "accommodates them as something fundamentally alien to the computer landscape."[67] As the title of her book suggests, Laurel argues that the theory and history of Western drama offer a number of ways for designers to incorporate an expanded awareness of human experience into their work. To make her case, Laurel draws comparisons between many of the concepts described in the HCI literature I have discussed and Aristotle's *Poetics*. Given the stated goals of creating pleasant, immersive experiences, design principles that incorporate an awareness of concepts like "pleasure and engagement are not only appropriate but attainable."[68] Laurel clarifies, however, that embracing a framework defined by drama theory would not mean abandoning the rigor of approaches grounded in scientific reasoning. On the contrary, by reaching back to and drawing on the Western critical and philosophical tradition, she claims that she is able to mesh logic and reason with discussions of the affective dimensions of human experience to present readers a more holistic approach to software design.

One advantage that drama offers over psychological frameworks, Laurel argues, is its explicit foregrounding of "immersive experiences." Whereas the psychological frameworks popular in HCI limit themselves to a focus on usability as a problem of human-machine communication and thus struggle to account for "what goes on in the real world, with all its fuzziness and loose ends," drama is able "to represent something that might go on, simplified for the purposes of logical and affective clarity."[69] The types of plots that drama provides are structured around producing highly meaningful experiences for audiences. The stage, as an environment for interpretation, creates these experiences by drawing on techniques like characterization, narrativization, and suspense that audiences are already very much familiar with and receptive to. Further, they are often willing to suspend disbelief in order to immerse themselves in dramatic experiences. Thus, like Shneiderman, Norman, and Winograd and Flores, Laurel calls on designers to acknowledge and exploit the arbitrary relationship between algorithm and interface. Because all representation of computation must be explicitly defined, there are few limits to how we can choose to represent software to users. However, Laurel goes a step further by arguing that there is no reason for us to constrain ourselves to reproducing "models of reality" if fiction offers designers a wider set of tools to develop compelling and engaging models of usability.

Laurel's principles of usability are, like those of other theorists I've discussed thus far, committed to keeping computation invisible. The sort of immersive experiences in drama that she believes should influence the design of personal computers are largely dependent on the ability of stagehands to perform their work invisibly: "Part of the technical 'magic' that supports the performance is embodied in the scenery and objects on the stage (windows that open and close; teacups that break); the rest happens in the backstage and 'wing' areas. . . . The magic is created by both people and machines, but who, what, and where they are do not matter to the audience."[70] Audiences do not want to know, nor should they be required to know, what happens backstage to experience the drama unfolding on stage. For audience members seeking an immersive experience, "the representation is all there is."[71] A computer's interface should function like a spotlight on a darkened stage, highlighting only what is necessary for the user to participate in the experiences constructed for them by the designer. A well-designed interface should present users with a distinct self-contained reality, with its own causal, narrative, and dramatic logics that function independently of the particulars of the computer system's functions. Anything that users encounter that is not understood as part of that self-contained reality, she concludes, interferes with their ability to suspend disbelief and so prevents them from realizing their desire for immersion within a digital space. Computational concepts have no place on the stage.

From this perspective, Laurel's dramatic principles of usability place designers in a position of total but ostensibly beneficial authority over user agency. Designers do more than develop software; they also decide what options are available to users in any given situation within a digital environment. Here, too, Laurel finds precedent in drama. This split between audience and the technical direction of a play is necessary for creating dramatic experiences. After all,

> what would it be really like if the audience marched up on the stage? They wouldn't know the script, for starters, and there would be a lot of awkward fumbling for context. Their clothes and skin would look funny under the lights. A state of panic would seize the actors as they attempted to improvise action that could incorporate the interlopers and still yield something that had dramatic integrity.

For Laurel, imposing limits on user awareness and agency is necessary to guide them toward the specific interpretations and behaviors they must accept as their own in order to have the experience the designer intends them to have. Without limits, user experience would "degenerate into a free-for-all, as performances of avant-garde interactive plays did in the 1960s." Yet, she insists, this narrowing of user agency and awareness does not relegate them to the role of merely a passive observer. Audience members do not need to join the actors on stage to participate in the performance. The interplay between the audience's responses to the action onstage and the actors' adjustment of their performance in light of those reactions makes audience members active participants in shaping the sense of immersion they share with one another. According to Laurel, they "become actors" themselves, playing their assigned role in the production.[72] While Laurel draws on drama theory to suggest that users influencing the performance of the actors is a form of active participation, it seems a far cry from the agency that actors or the director possess.

Laurel further develops her principles by explaining that narrow definitions of user agency are essential if users are to have any sense of agency at all in a digital environment. If designers limit user awareness and disallow certain types of behavior, users will be better able to see "causal relations among events [that] are not obscured by . . . randomness and noise."[73] Giving users freedom to do as they please is thus less important than providing them with a system that presents a clear and readily understandable logic so that they can interact with it immediately while also knowing what the effects of their actions will be. Ideally, users should experience constraints that limit "not what [they] can do, but what [they] are likely to think of doing."[74] What is left unspoken here is that no such constraints exist for software developers. Although Laurel praises computers as a new medium capable of making possible as yet unimagined forms of creativity, it is clear that within her framework only the designer as director enjoys such freedom. Users experience computing through a more limited awareness provided to them that continually redirects their attention away from the technical aspects of computation. They can only play the parts written for them. Laurel's principles of usability may not assume a universal theory of cognition, but her insistence on limiting the user's ability to engage with computation naturalizes transparent design as an intuitive solution to the problem of creating universally usable and useful interfaces. Laurel may claim in her rhetoric of usability that she is proposing a new approach to design that is based on ideas from the arts and humanities, but with respect to the relationship between designer and user, her principles of usability very much resemble those common among early HCI theorists.

The structural similarities between Laurel's principles of usability and those of Shneiderman, Norman, and Winograd and Flores are especially evident when one compares her work to Suchman's. Like each of the theorists addressed in this chapter so far, Suchman argues in her early research on usability that theories of human information processing represent an intentionally constrained, artificial model of human experience. She also cites Dreyfus's work as a formative influence on her understanding of usability in her 1987 book, *Plans and Situated Actions*. Whereas the rhetorics of usability of each of the theorists I have considered up to this point approach the sociocultural complexities surrounding usability as problems to be solved, Suchman's frames them as starting points for extended explorations of technologically mediated communication. For Suchman, computers are not something that designers construct for users to interact with but a tool through which users and designers interact asymmetrically. Her criticism of approaches to usability that incorporate human information processing serves primarily to help her establish principles for the ethnographic observation of users as part of an ongoing process of inclusive design.

Suchman begins by asserting that the questions that have troubled HCI researchers are not new questions to anthropologists, particularly those like herself who adopt an ethnomethodological framework. In particular, Suchman sees the problem of "mutual intelligibility" as a key concern for usability:

> The notion that we act in response to an objectively given world is replaced by the assumption that our everyday social practices render the world publicly available and mutually intelligible. . . . The outstanding question for social science, therefore, is not whether social facts are objectively grounded, but how that objective grounding is accomplished. Objectivity is a product of system practices, or members' methods for rendering our unique experience and relative circumstances mutually intelligible.[75]

Referring to the work of C. S. Peirce and Harold Garfinkel, she explains that because language is indexical—in the sense that words and other linguistic signs refer to ideas or phenomena in the world—acts of interpretation rely on our experience of context.[76] Our sense of a stable world is not the product of an objective reality that can be experienced uniformly by all participants or a shared body of meanings but rather of "our tacit use of the documentary method of interpretation to find the coherence of situations and actions" by searching for "uniformities that underlie unique appearances."[77] Theories of human information processing assume that humans, on some level, follow a universally shared set of rules for interpreting and making decisions about the world around them, and so two people experiencing the same situation should have a very similar understanding of it. Yet, Suchman notes, even were we to accept that human interpretation of the world is based upon universally shared cognitive principles, then we would still need to account for the fact that the interpretations we construct are mediated by languages, behaviors, and experiences that are not universally shared. Specific shared understandings thus cannot be assumed. They are not natural and take considerable effort to build and maintain.

Even the appearance of shared or similar interpretations does not itself suggest that objective interpretation is possible, nor even that a common or universally shared set of rules for interpretation exists. Suchman explains that it is not uncommon for two people to recognize through casual observation a particular causal logic—whether understood as physical laws, social norms, or common methods of cultural experience—but provide very different explanations of that logic. For example, two people may assume that the proper response to a sneeze is to say "bless you" but give two distinct reasons as to what purpose it serves and why that particular response is more appropriate than others. Here, Suchman implies that usability should be approached as a cultural rather than a cognitive concern. By introducing the idea of mutual intelligibility to usability, Suchman highlights how early HCI researchers were more concerned with developing frameworks to inculcate specific behaviors than with helping users learn to understand the logic behind those behaviors. Approaches that normalize behavior establish a boundary between designers and users that makes it increasingly difficult to undertake the work of developing a shared understanding between them.[78] Suchman shares with the other theorists discussed in this chapter a belief that principles of usability based on human information processing do not account for the fullness of human experience; however, the principles of usability she develops out of that criticism sharply contrast with theirs. Owing to the role of mutual intelligibility in usability, she maintains, the tension between computation and culture cannot simply be resolved through good design but must instead be continually negotiated.

Shifting usability farther away from a focus on cognition also troubles the idea that all computer use is necessarily driven by planned actions. Theories of usability based on human information processing, Suchman explains, assume that before taking any action, humans explicitly define a goal and then proceed to map out a plan of action, following each predefined step until a conclusion is reached or the sequence is interrupted. As I have shown, this is a defining principle of frameworks like GOMS, particularly in the pseudocode its practitioners develop to simulate behavior. It is also visible in Shneiderman's and Norman's work through their discussions of semantic knowledge, syntactic knowledge, and mental models. Yet Suchman notes that most of the tasks we successfully undertake do not require us to first enumerate a complete series of actions. We may consciously decide to pursue a goal, like going to the grocery store, and even subgoals, like purchasing ingredients for a recipe we would like to prepare that evening, but we do not explicitly plan each action we take in pursuit of those goals before acting. We do not plot the exact path we will take on our hypothetical trip to the grocery store, nor do we need to memorize the exact location of items, calculate the precise number of steps we will make within the store, or predetermine the specific angles from which we will examine vegetables before deciding whether or not to put them in our basket. We make these movements spontaneously, without the need for them to be defined in exact measurements, and often in response to changing circumstances in our immediate situation. Like Winograd and Flores, Suchman asserts, drawing on Dreyfus's commentaries of Heidegger, that we often "construct rational accounts" of our behavior "before and after the fact," but "when action is proceeding smoothly, it is essentially transparent to us."[79] Plans are thus best understood as abstract representations of "situated actions," ones made in the moment and ever changing in response to a fluid, tacit understanding of our environment.

Software designers, Suchman continues, must recognize that users continually act on incomplete but functional understandings both of the world around them and of their own actions. Paradoxically, "the efficiency of plans as representations comes precisely from the fact that they do not represent those practices and circumstances in all of their concrete details. . . . It is frequently only on acting in a present situation that its possibilities become clear." Often, we "do not know ahead of time, or at least not with any specificity, what future state we desire to bring about."[80] Again, Suchman here appears to develop a principle shared by Shneiderman, Winograd and Flores, and Laurel. Yet the frameworks proposed by these other theorists ultimately impose a goal-directed structure onto user behavior via constraints. The principles of usability defined by Shneiderman, Norman, Winograd and Flores, and Laurel all seek to reduce uncertainty among users about how to proceed at any given point by narrowing the possibilities for action so that a destination and clear path toward it becomes intuitively clear, as if users chose it themselves. But, as Suchman might respond, this approach only ensures that users will follow a path, not that they will recognize why the path is shaped a certain way, nor necessarily be able to anticipate where it or similar paths may lead. A transparency of interaction will hide the structures that guide users' actions, making them believe in hindsight that the intended course of action modeled by the interface for them was one that they chose freely themselves. Similar results across users can be socially engineered but they do not necessarily reflect a shared understanding of how the technology works nor do they ensure that users will necessarily understand how their actions produced a specific result.

Although Suchman concludes that we should approach usability as a project of communicating with users about technological systems, she acknowledges that this perspective is complicated by the fact that human-to-computer communication is asymmetrical. Mutual intelligibility is a problem that sociologists have devoted a great deal of time and energy to documenting and explaining. How we reach a point of mutual intelligibility may be hard to understand from a theoretical perspective, but in human-to-human communication it is something "we achieve in our everyday interactions[,] sometimes with apparent effortlessness, sometimes with obvious travail," but always as "the product of *in situ*, collaborative work."[81] Human-to-human communication can be goal directed in the sense that we tend to have an abstract reason for striking up a conversation with someone, but the course of a conversation is not typically preplanned. Instead, communication is usually experienced as transparent. During conversations, we make interpretive judgments continually, often without pausing to reflect on them, intuitively ascribing intention and meaning to the other participant's words based on our individual awareness of the situation while relying on our experience with similar, prior situations in order to disambiguate between a variety of potential meanings. These are complex interpretive actions, as evidenced by the social tension that misunderstandings can produce, but most of the time we navigate this complexity seemingly effortlessly. Even in cases when communication breaks down, human beings rely on a number of conversational strategies that they draw upon without thinking about it, without removing themselves cognitively from the normal flow of communication. In human-to-machine communication, however, computer software only responds to the specific patterns of user input that designers planned for and anticipated. As a result it is "not simply that communicative troubles arise [in human-to-machine communication] that do not arise in human communication, but rather that when the inevitable troubles do arise, there are not the same resources available for their detection and repair."[82] When human-to-machine communication breaks down, the burden is placed entirely on the user to reflect on and disambiguate the process. Here we can see how many of the same principles proposed by other HCI theorists I have discussed seem advantageous. They can ease the burden on the user by constraining the range of things that must be reflected on to disambiguate the situation.

However, Suchman takes a different path forward. Usability theorists, she observes, seem aware of the asymmetrical nature of human-machine communication but appear more interested in exploiting rather than exploring it. So long as designers view usability as a way to pleasantly conform users to an idealized model of knowing and doing, "machine-interaction with the world, and with people in particular, will be limited by the intentions of their designers and their ability to anticipate and constrain the user's actions."[83] In other words, while the experience of transparency in human-to-human communication results from the practiced effortlessness of interpretation, the experience of transparency in human-machine communication has thus far only been realized via a designer's ability to predict users' interpretive behaviors and use those predictions to passively move them to accept the actions they take in a technological environment as freely and intuitively undertaken. Users experience software as transparent only insofar as they do not realize the way it moves them to take specific sequences of actions. Thus, Suchman suggests that the pursuit of transparency in human-machine communication will inevitably prioritize normative relationships between designer and user. For many of the other theorists covered in this chapter, that privileging of operation over understanding is the point. For Suchman, however, an approach to design that focuses on teaching users intended behaviors and not helping them learn to understand a given software system has significant negative consequences for designers, as they will inevitably find themselves solving the same kinds of problems over and over again with each new technology.

And it is on this point that Suchman's approach is distinct from Laurel's. Both propose principles of usability that are defined through an engagement with theories and methods from outside of cognitive psychology and engineering. Whereas Laurel assumes that designers can ensure a relatively singular experience among users by constraining them within narrowly defined, immersive systems of logic and affect, Suchman asserts that no two users will ever share the exact same experiences with and understandings of technology, even if they are able to produce similar results when using it. Rather than proposing principles of usability that smooth over potential uncertainties about the identities, interests, or learned behaviors of users, Suchman concludes her study by calling on designers to see uncertainty as a productive opportunity. There may indeed be a tension between computation and culture, but Suchman suggests that this conflict persists because technologists continue to develop principles of usability that provide designers with new and innovative ways to avoid directly engaging with users and with novel explanations as to why users should not participate more substantially in the design process. Software design is a discipline that should encourage constant dialog with users. We must set aside the idea that humans are communicating with machines and recognize that software design is about humans communicating with each other through the interface.

Designers are continually sending messages to users. The asymmetrical nature of design does not change this fact. Such asymmetry does mean, however, that if transparency is the goal then designers must anticipate all possible messages that users may try to send them and precompose all possible responses needed to help address the ambiguities that will inevitably arise. When described in this way, transparent design is not simple at all.

Conclusion

The way that software designers leverage a concern for users to justify their push to normalize personal computer users and their contexts of use is a recurrent theme in Suchman's subsequent work, especially in her efforts to contribute more directly to the fields of science and technology studies and feminist science studies. She later clarifies her view that usability remains focused on a problematic political relationship between designers and users. Principles of transparent design maintain two sets of boundaries: the first between professional design and technology's local configuration and the second between the technical expertise of designers and users.[84] In addition to propping up the myth of the "lone (male) creator of new technology," these principles lead designers to rely on a series of "surrogates, proxies, stand-in's [*sic*], for 'the user,' designed to allow the creation of usable technologies in the absence" of users. It is not merely an unintended consequence of this approach to design that leaves users with no agency other than what designers afford them. Rather, Suchman suggests, designers are rewarded by the professional culture they are immersed in when they maintain these boundaries and thus seek to construct models of usability that minimize the ability of users to make decisions for themselves even during implementation.[85] While Suchman's *Plans and Situated Actions* gained attention within HCI and helped cultivate some ethnographic practices in usability research, Suchman believes that designers have misinterpreted her basic argument. Twenty years after its initial publication, she clarified in the book's second edition that usability theory must "abandon the attempt to create the 'self-explanatory' machine in favor of encouraging social arrangements that provide for the necessary time and resources needed to incorporate unfamiliar artifacts effectively into relevant forms of practice."[86] Rather than read her work as a way to negotiate the sociocultural complexities of design, she observes, software developers have instead embraced the normative potential she warned about. To move forward, we must insist like Suchman that designers take steps to ease the asymmetry between themselves and users by allowing users to talk back, whether that be through an increased ability to reconfigure a technology in novel ways during implementation or through more direct involvement in the design process itself.

To be clear, like Dreyfus, I do not mean to dismiss designer's technical expertise in offering a critique of usability. Rather, my focus has been on the way that a concern for users is leveraged to justify claims of cultural and social authority by technologists. It is in some ways miraculous that the increasingly complicated systems that technologists build manage to work as well and as consistently as they do. But as I have shown in this chapter, it is important that we recognize that although for nearly forty years HCI theory has leveraged rhetoric from the humanities to suggest that usability engages in similar modes of inquiry and that its abstract appeals to universal use are responsive to the complex needs and identities of computer users, this is not truly the case. These same moves are present in contemporary discourse, even as the specific terms and tropes used by Big Tech and start-ups alike have been updated and refined: identify a problem as the result of an older technology that is disengaged from some dimension of culture or society; recognize the complexity of the problem; and, finally, propose a new technology that is represented as being able to provide a simpler, more universally usable solution that everyone can benefit from. In popular accounts of tech culture, these moves are portrayed as commonplace but hollow gestures, ones intended to persuade venture capitalists but not ones that reflect genuine commitments by their rhetors.[87] Two important questions that we must consider are why this problematic leveraging of humanistic rhetoric and concerns persist, even as many at least tacitly recognize them as a pretext for design rather than its primary goals, and what an alternative might look like.

Were we to focus on the commercialized nature of personal computing, we could conclude that the persistence of these seemingly culturally engaged rhetorics of usability owes to their commercial success both with investors and with consumers. But there are other frameworks available both in HCI and in other fields that continue to critique normative practices in design and the privileging of expediency over users' needs by adopting concepts and modes of inquiry from the humanities. Consider, for example, Jeffrey Bardzell and Shaowen Bardzell's recent work on "humanistic HCI." Arguing that software designers would benefit from the reflective and interpretive practices modeled by literary criticism and acknowledging that HCI is often normative in practice despite rhetorics of usability that promise liberation, they call on humanists to join the field of HCI in order to help researchers "live up to their own utilitarian, ethical, and aesthetic ideals."[88] Practicing humanistic inquiry within HCI would transform the designer from an applier of objective principles into an "expert subject" who through debate and conversation with peers is able to draw on a broad range of critical and interpretive theories and thereby illuminate HCI's blind spots and refocus design practice around the field's stated ideological and epistemological commitments.[89] Now, contrast their principles with those proposed in frameworks like design justice or human-centered design as described by Sasha Costanza-Chock and Natasha N. Jones, respectively.[90] Both highlight the importance of a self-critical perspective in design, one that recognizes how designers and users are often in conflict with one another in ways that result in the silencing of the latter. Rather than assume that designers, if they work with the proper values in mind, can develop models of universal usability, these frameworks instead insist on making a space for users to propose and explore alternative definitions of what counts as usable and useful.

One reason why theorists continue to produce rhetorics of usability that invoke humanistic values may be that the critiques they raise are still valid even if the principles they propose are problematic. Stepping outside of the American context that has been the focus of this book, we can see some evidence of this broader support for a more socioculturally engaged approach to design. In Scandinavia, for example, a workplace democracy movement has centered on what is called "participatory design," its goal being "to enable those who will use the technology to have a voice in its design, without needing to speak the language of professional technology design."[91] While the participatory design movement developed over the course of the 1960s and 1970s, overviews of the field observe that advocates have responded favorably to early work in HCI, particularly Winograd and Flores's centering of human experience and Suchman's recognition that design shapes rather than seamlessly fits into that experience.[92] However, participatory design has its own problematic politics. Being situated firmly in the workplace, as Constanza-Chock notes, its practice is often strongly influenced by existing office politics. Workers may not feel comfortable expressing their views about technology for fear of reprisal from management. And even when they do express their views, the design process may still be extractive in the sense that their input may serve only to make design decisions that benefit management seem more palpable to users (as I have noted, this is a scenario that Winograd and Flores suggest is an advantage of transparent design).[93] Nonetheless, participatory design does at least foreground the types of difficult cultural, social, and political questions that user-friendliness and transparent design forgo in favor of immediate, pleasant, and immersive user experiences.

In the conclusion that follows this chapter, I consider what must be done to get out from under the politics of user-friendliness. A core assumption visible in HCI literature and present throughout the rhetorical history that I have traced in previous chapters is that users do not have the ability or the desire to wrestle with the technical complexities of personal computing. So long as this assumption holds sway, whether expressed in marketing language or through cognitive theories, the boundaries that Suchman identifies will remain in place. Breaking them down so that we can reject the idea that computational culture must be curated for us by elite innovators will require us to promote a counter rhetoric of usability that values complex engagement and that does not position friendly, simple interactions as the only acceptable or natural goal of design.

Coda

Imagining an Unfriendly Future

My goal in this book has been to follow several of the threads in the rhetorical history of usability in personal computing that have been entwined as "user-friendliness." These threads have included how the concept of friendliness implies a separation between computation and culture, how the narrative framing of personal computing as countercultural supports technological expediency, how the continued pursuit of more transparency has been driven by fear of a computer literacy crisis, and how a concern for users can serve to flatten the sociocultural dimensions of computing. Each chapter has focused on at least one of these threads, documenting their emergence and the roles they have played in shaping our relationship to digital culture. In this conclusion, I pull these threads back together in order to imagine an "unfriendly" future that explicitly makes room for users to critique and intervene in computation rather than leave it to be managed from afar by benevolent designers. While user-friendly approaches to design are often portrayed as a way to define usableness and usefulness in response to users' needs, I have shown that the idea of friendliness has been invoked historically to justify normative approaches to design. Personal computing technologies are socially constructed; however, designers and users have distinctly different degrees of agency when it comes to influencing that construction. Since the 1980s, major hardware manufacturers and software developers have devised rhetorics of user-friendliness and models of transparent design to structure user agency in ways that privilege their own economic, political, and cultural goals over those of users. Users can reject their rhetorics and even propose counter rhetorics in order to explore new forms of use. But, as I have shown, rhetorics and models of usability often work together in complex ways that serve to naturalize ideal understandings of computation and intended use behaviors.

If we are to imagine an unfriendly future, we must begin by recognizing that user-friendliness helps designers to conceal not just the technical aspects of computation but also their constant interventions in digital culture. One lingering concern that I have had as I have reflected on my book is whether or not my articulation of the politics of user-friendliness reads as determinist even as I insist that personal computing is socially constructed. These understandings of technology are not necessarily mutually exclusive. As Paul Edwards has noted in his work on infrastructure studies, there is a momentum that can set in when technologies establish standards such that later technologies become dependent both on the material functions of earlier ones and the social structures that have developed around them. These dependencies constrain our social construction of new technologies and impose limits on our ability to later repurpose them toward new ends.[1] Applying this lens to personal computing, we can see that consumers have not only come to accept user-friendliness and transparent design but also to expect it. It is important here to remember Annette Vee's description of literacy as sociomaterial. Just as programmers learn to operationalize the phenomena they model as algorithms and data structures via the methods and conceptual frameworks associated with specific programming languages, so too do users learn to think about how computers work and what roles computers should play in our lives via the specific rhetorics and models of usability presented to them.[2] User-friendliness and transparent design have become so ubiquitous that it is difficult for us to imagine alternative relationships not just to digital culture but all of culture, given how thoroughly computers have been integrated into our everyday lives.

Moreover, the naturalization of user-friendliness is no accident. One key design norm discussed in most usability handbooks is "consistency." In general handbooks, this principle is presented matter-of-factly as a framework that helps users more immediately recognize the use behaviors designers intend them to adopt by emphasizing conceptual similarities between homologous tasks. But in handbooks written by platform developers like Apple, it becomes a mandate to ensure that all software is understood through the language and behaviors modeled by Apple's operating systems.[3] When all software behaves similarly, its design norms begin to be treated akin to natural laws. Imagining an unfriendly future requires us to step back and reconsider the political implications of seemingly unobjectionable design principles so that we can recognize how they often do not merely support ease of use but also serve the interests of the elite technology companies that are able to define the norms of personal computing.

In the friendly future we inhabit, software designers often structure our relationship to software to maximize their momentum and minimize our agency so that we experience designers' constant interventions in our lives as inevitable and unavoidable. While journalists were abuzz in the late 2010s about a coming "techlash," we have continued in our everyday practice to accept Big Tech's constant revisions to the software we rely on at work and at home. It is rare that a day passes without our devices downloading some new update that slightly alters some aspect of our digital agency. If someone were to enter our offices at night and rearrange one or two of the items on our desks, we would feel violated. But when the equivalent happens on our phones, laptops, and gaming consoles, we have been enculturated to accept it as somehow necessary and move on. To reclaim our agency, we must stop and ask why our hardware and software must change. Today, computer crud is wrapped in a rhetoric of user-friendliness: we are told that the surveillance and the continuous stream of updates we are subject to are somehow necessary to improve user experience. For whom is it improved? And what was wrong with it before? Pushing back against the normative power of user-friendliness is difficult, in no small part because its rhetorics of usability have positioned transparent design as a technological realization of Western democratic ideals. The associated principles of universal usability, immediate access, and personal empowerment through simplification have been embraced not only across the American political spectrum but also increasingly transnationally.[4] In order to imagine an unfriendly future, we need to challenge the idea that computers must simply be "easy to use" and consider other usability goals. While digital media have changed both discursively and materially in the United States since the 1970s and 1980s, two key principles have endured almost unchanged: that computers are "personal" and that transparency can "come for free." If we are to embrace an unfriendly future, we will have to reject these principles.

Computing Has Never Been "Personal"

The idea of "personal" computing may have begun as a marketing gimmick, but it has come to inform our desire for computing power: something that will let us cut through the complex challenges in our lives and on the job, empowering us to succeed in any context. Influential popular histories of the period have helped to establish a narrative framework through which we understand personal computing as an effort to radically decentralize computation in order to produce tools for self-empowerment. If we devoted ourselves to the self-study of computation through new, individualized models of usability, this narrative implies, we would become as gods and take control of our relationship to the world around us. But the radical decentralization that Silicon Valley's mythical hobbyists pursued was fleeting. Within just over a year, the widespread adoption of Microsoft BASIC had begun and was largely accepted because it was more technologically expedient than the more convivial alternatives. Appliance computers would push further for recentralization around themselves. Not only did many appliance computers eventually incorporate Microsoft BASIC, but their manufacturers also leveraged proprietary designs and withheld technical information about those designs in order to exert control over early digital media ecologies. This process has continued and even accelerated into the present. Cloud computing is the future now. But beneath all the marketing buzz, it looks very similar to the older models of centralization that personal computing was supposed to do away with. Very little about our computing devices and practices today is uniquely our own, and yet personalness is still a key idea in today's digital culture. Even in an age of connectedness, we still are encouraged to think of the digital environments displayed in our hands, on our desks, or through our televisions as private to us. Even as we venture out from them into online communities, media streams, virtual worlds, and online marketplaces, we are still encouraged to see many aspects of them as "personal" to us: home screens, profile pages, and desktops that we can customize and fill with our own data. Despite the many changes to digital culture since the 1980s, our user-friendly devices continue to promise us that the more we invest of ourselves in them, the more usable and useful they will become.

Rejecting personalness can help expand our understanding of usability in several important ways. First, it can help us to expose and confront the biases that user-friendliness's claim of universal usability obfuscates. During the first decade of personal computing in the United States, hobbyists, journalists, software designers, industry executives, and usability theorists characterized personal computing as a technology that everyone would be able to participle in. Their gestures toward increased access or recognition of a diversity of user identities would imply in a generous reading a commitment to equity through design. Yet the systemic issues they claim to address are never defined, nor is any specific strategy for acknowledging difference and incorporating other voices offered. As part of her discussion of the "New Jim Code," Ruha Benjamin explains that "innovations reflect the priorities and concerns of those who frame the problems to be solved. . . . A narrow investment in technical innovation necessarily displaces a broader set of social interests. This is more than a glitch. It is a form of exclusion and subordination built into the ways in which priorities are established and solutions defined in the tech industry."[5] In other words, universally usable technologies do not produce universally shared experiences. Rather, the promise that every user can establish a personal relationship with computing pushes users to align themselves with the values and interests of a technology's designers. That alignment is a condition of use, a cost for access that can be negligible if you are privileged enough to resemble the type of user designers had in mind and that can be quite high if you are not. Personalness suggests that we can fit technologies into our lives as we see fit, but in practice the goal of user-friendliness is to nudge us to tacitly accept as our own the interests of a technology's designers. Efforts to design for accessibility have helped to make a space for considering different embodied identities; however, critics have noted that this work can also be performed with a similarly normative purpose, folding those who are "over there" into the norm rather than valuing alternative models of usability.[6] Rejecting the rhetoric of personalness can thus help us to expose and confront the biases that are necessarily embedded in all products of human creativity but that are sublimated by user-friendliness.

Additionally, it is important to recognize how appeals to personalness leave us subject to the managerial logic enacted by user-friendly technologies. As I have shown, rhetorics of user-friendliness have historically emerged out of corporate computing contexts. Advances in usability were often understood to benefit workers first; home users were more of an afterthought. Transparent design conceals this managerial logic from us, just as it conceals the merely technical aspects of computation. Personalness encourages us to focus on how valuable computers are to us and to brush aside the ways they make us valuable to others. Shoshana Zuboff has written extensively about the way that personal computers have been restructured to support a new sector of the American economy devoted to the unseen extraction, packaging, and resale of personal data. Zuboff focuses mostly on internet-based, social media technologies, but user-friendliness supports in more mundane software genres the same practices she describes in social media.[7] Operating systems and web browsers are not software we can ignore or avoid, as they mediate nearly every aspect of our digital agency. Their now always online nature means that designers and their managerial logics are now always invisibly present. While working on the manuscript for this book on my home computer, for example, I was greeted one morning by an email from Microsoft regarding new weekly "insights" about how I use Office applications and their Outlook email service. Microsoft notified me that it was collecting and analyzing data about my use of its software. It was a rare moment when the veil of transparency was lifted, ostensibly under the guise that these insights would help me manage my time and thereby "bring harmony to work and life."[8] I disabled the feature, but I am sure that Microsoft is still collecting data about how I use its productivity suite. There is no way for me to know what simply toggling that switch actually accomplished. Because of the arbitrary relationship between interface and algorithm, all I can know for sure is that I no longer get a weekly email from Microsoft.

Personalness remains critical to a major emerging area of concern in usability research: trust. In our ultra-transparent, user-friendly future, our lives are wrapped up in a proliferation of automated systems, with new ones being developed and commercialized all the time. Trust is quickly becoming a foundational principle for design, as users begin more and more to feel the effects of a proliferation of automation and ask themselves whether these unseen mechanisms really are acting in our best interests.[9] However, when user-friendliness and transparent design are working in sync with one another, it is very difficult to know for certain whether an application is doing everything that its interface claims, something more, or something different entirely.[10] If this possibility seems overblown or conspiratorial, then take a moment to step outside of the bland construction of the universal user. Vulnerable populations who are put at risk by violations of trust cannot ignore this possibility. What should we say to users of a Muslim prayer app whose data is unknowingly shared with a company contracted to identify terrorists?[11] Or to women tracking their reproductive health using an app that quietly sells their data to advertisers without asking for their consent?[12] Should we simply tell them to be more responsible about the software they choose to install? William Gibson imagined that we would need to pry personal computers out of the hands of the big corporations who hoarded technological power for themselves.[13] But we do not live in that future. We live in a user-friendly one where Big Tech companies like Apple, Microsoft, and Google and smaller, unknown developers alike give away their technologies. Their rhetorics of usability draw on narrative framings co-opted from the arts and humanities to suggest that we can use technology to reflect on ourselves and enrich our lives by enhancing our relationship to culture and society. A sense of personalness encourages us to ask simply what we want to know or do and then trust that the right app will fulfill its promise to deliver our desires to us.

Transparency Does Not "Come for Free"

Rhetorics of user-friendliness promise that personal computing technologies can be both phenomenologically transparent and transparent to human intention. The models of usability they describe can be integrated seamlessly into our lives, invisibly supporting but not intervening in our ongoing sociocultural activities. This goal is impossible, of course. While user-friendly technologies may fade from our awareness at times during use, they necessarily intervene in the activities they are integrated into by virtue of the decisions made elsewhere, on our behalf, and the resulting automations that structure those activities. The relationship between designer and users not need be predatory or exploitative; however, that has all too often been the case, and rhetorics of usability push us to overlook or accept that imbalance of power as a good and necessary thing. As I have noted, user-friendliness naturalizes transparent design by implying that computation is inherently in conflict with culture. Whereas the hobbyists' mythos represented the conflict as being between technocracy and democracy, user-friendliness portrays computing as potentially demanding too much of our time and attention. Designers posit that computation exists outside of culture, in a space they can control, and they promise to contain it by minimizing its demands on us.[14] This claim calls to mind Ivan Illich's point that over time elite, technoscientific institutions become primarily self-serving. The apparent grayness or banality of personal computing drives their insistence that paying attention to computation is not worth our time. Computation is a neutral domain, full of merely technical concerns that have little direct relation to the cultural and social concerns that define our lived experience. Transparency is thus believed to "come for free" because there is no cost to users in hiding the technical aspects of computation from us.

At the same time that we are told that the technical aspects of computation occupy a space removed from culture, Silicon Valley has also championed a belief that technological innovation can address even the most complex sociocultural concerns. Big Tech has positioned itself as uniquely suited to harness technology to address the most difficult challenges facing the world today. And in the meantime, it can take care of smaller nuisances, too. Silicon Valley's elite innovators promise that the Gordian knot that wicked problems present can be sliced through with the right technology, even and especially if those problems are caused by existing technologies. The proliferation of transparent, user-friendly technologies atop the ever-growing, interlocking information systems we interact with daily makes it difficult for us to agree on the precise nature of these problems, let alone devise effective strategies for addressing them. Insisting that the merely technical aspects of personal computer systems should remain secret from us does come with a cost, specifically the inequities in computer literacy that user-friendliness encourages between designers and users. By treating the hidden aspects of personal computer systems as merely technical, Big Tech recreates the very politics that the hobbyists' narrative was directed against. Elite software developers refuse to acknowledge the cultural, social, and political influences on their work by insisting that the problems they are trying to solve are merely technical, only to turn around and moments later claim that the technologies they are building will change the world. By keeping the complexities of the software they build hidden from us, they hinder public examination of those influences.

Thus far, the more visible calls to examine the cultural, social, and political biases in technical systems have been advanced through a libertarian political framework that frames any attempt at policy-driven intervention as either introducing unnecessary further complication or a challenge to entrepreneurial freedom. In this way, the free and open source software movements are essentially recreations of Stewart Brand's New Communalist movement. They have inspired and sustained significant separatist communities defined by alternative design principles. But the practices of code sharing they encourage have done little to change the politics of Big Tech, and many of their separatist communities have a reputation for being hostile to people who cannot demonstrate a satisfactory level of technological expertise. Major software companies have at times responded directly to public concerns about their management of hidden systems. However, they have done so primarily by introducing new mechanisms for personal accountability, using the same rhetorics and models of usability that got us into this mess to suggest that we can, individually, take control over our relationship to the systems transparent design keeps hidden from us.[15] To date, Big Tech companies have been held accountable only by public outcry and even then only if that outcry translates to a significant threat to shareholder value. Digital freedom activists see these concessions as victories but are largely content to allow Big Tech to govern itself and comparatively more tolerant of surveillance when it is conducted by private companies rather than government actors.[16] As Illich reminds us, the sort of conviviality that Silicon Valley teases us with is only possible through public institutions dedicated to preserving individual autonomy. In the long term, the most important way to realize an unfriendly future is to recognize that technology must be publicly regulated via formal mechanisms that ensure political transparency rather than transparent design. Government systems may not be perfect, but the continued belief that any form of regulation is a threat to digital culture is a reminder that user-friendliness still serves a critical role in supporting the technological expediency that has driven the personal computing industry since the late 1970s.

In the shorter term, the moments when the boundary between designers and users falls away provide an opportunity to explore the costs of transparent design. As Lucy Suchman has observed, software developers often treat design as everything that occurs prior to implementation, when a technology is released to users.[17] Good design ensures that users cannot change the technology and will implement it as intended. Designers have normalized user behavior through rhetorics of usability since the 1970s, but today's always online models of usability are even more insidious, affording designers the ability to police user behavior after release through patches, hotfixes, and updates. However, some forms of software available today are designed with an understanding that users will want to venture out from those intended behaviors. One of the best examples in recent memory is the Myspace social networking platform. In its earlier iterations, the ability for Myspace users to customize their profile by defining layouts in HTML and CSS introduced many users to computational concepts. It also served to reveal these merely technical concerns as having cultural and social dimensions, as users communicated through them by viewing, copying, and modifying their code. Similarly, web browsers have allowed add-ons for decades now, helping users to extend the functionality of their software in ways the original designers never anticipated. Additionally, the "modding" communities that have formed around video games since the DOS-era have helped erase the boundary between designer and user by transforming game engines into tools for creative expression and for developing a form of computer literacy that engages more directly with computational concepts. Even if designers can only be motivated by financial self-interest, then it is important to recognize, too, that modification has also helped to sustain user interest in, and sales of, those games that support it for many years after release.[18] More importantly, permitting user modification foregrounds the communicative nature of personal computing. Every time we sit down at our computers, we are not simply interacting with algorithmic constructs but with the designers who built them. Transparency hides their presence from us, but being permitted to modify our software can help us to better recognize how they intervene in every action we take in digital environments. Permitting modification expands user agency and helps us learn to talk back to designers more effectively instead of leaving us to choose between simply responding in approved ways or totally rejecting a technology.

In addition to distorting our perception of the relationship between designers and users, transparency also hides the social costs of technology. User-friendly interfaces discourage reflection, pushing us not only to look past how they work but also how they were made. Pulling back the veil of transparency is a first, important step to forcing public accountability for the exploitative practices that produce our technology.[19] Tech workers stand to benefit from an unfriendly future too. While I have criticized designers and developers in this book, I am more broadly referring to the corporate entities that drive design and in whose interest final negotiations are made between the definition of user needs, a technology's intended use behaviors, and how its model of usability will fit into a financial strategy. The same principles of abstraction and encapsulation that make up transparent design can similarly be leveraged to manage individual programmers in ways that keep their focus on intellectually fulfilling problems while also keeping secret from them information about how the code they write will be implemented later. Many Google employees, for example, only discovered that the machine learning and artificial intelligence problems they were solving were going to be applied to a censored search engine after news about the project was leaked to the press.[20] In these moments, an inverted version of Suchman's boundary comes into play for tech workers. For this reason, tech labor unions will likely be a key ally for users as we seek an unfriendly future.

Transparent interfaces ensure that there is always a limit to what we can know about software, not just how it works, but how it was made, by whom, and for what purpose. Lifting the veil of transparency will not simply make the sociocultural aspects of software wholly knowable to us. Since the late 1960s, its ever-increasing scale and scope has proven challenging to comprehend even for professional developers.[21] But in an unfriendly future, its mechanisms would be exposed and recognized as being more than merely technical. Computers could still be "easy to use" but our understanding of what counts as "easy" would change in recognition of the costs and conflicts at stake in our use. Further, by avoiding universal models of usability, we could pursue models that would support different definitions of user agency rather than facilitate normalization. Computers would not be "personal," but they would not be managed in secret by benevolent oligarchs either. An unfriendly future would not erase all of the inequities around computing, but it would foreground that they need complicated social responses, not simplifying innovative solutions. If user-friendliness encourages designers to simplify our engagements with computation, an unfriendly future is one in which complexity is valued and confronted.

Notes

Introduction. The Politics of User-Friendliness

1. "Steve Jobs on the Future of Apple Computer," *Personal Computing*, April 1984, 243.

2. Walter Isaacson, *Steve Jobs* (New York: Simon and Schuster, 2011), 132.

3. Eric Z. Goodnight, "Macs Don't Make You Creative! So Why Do Artists Really Love Apple?," How-To Geek, March 9, 2011, https://www.howtogeek.com/howto/45498/macs-dont-make-you-creative-so-why-do-artists-really-love-apple/.

4. Susan Gasson, "Human-Centered vs. User-Centered Approaches to Information System Design." *Journal of Information Technology Theory and Application* 5, no. 2 (2003): 29–46; Julie A. Jacko, ed., *Human-Computer Interaction Handbook: Fundamentals, Evolving Technologies, and Emerging Applications*, 3rd ed. (Boca Raton, FL: CRC Press, 2012); Jeff Rubin and Dana Chisnell, *Handbook of Usability Testing: How to Plan, Design, and Conduct Effective Tests*, 2nd ed. (Indianapolis, IN: Wiley, 2008).

5. Todd Bishop, "How Valve Experiments with the Economics of Video Games," GeekWire, October 23, 2011, https://www.geekwire.com/2011/experiments-video-game-economics-valves-gabe-newell.

6. Michael S. Mahoney, "Issues in the History of Computing," in *History of Programming Languages*, vol. 2, ed. Thomas J. Bergin Jr. and Richard G. Gibson Jr. (New York: ACM Press, 1996), 772–81; Nathan Ensmenger, "The Digital Construction of Technology: Rethinking the History of Computers in Society," *Technology and Culture* 53 (2012): 753–76.

7. Paul E. Ceruzzi, *A History of Modern Computing*, 2nd ed. (Cambridge, MA: MIT Press, 2003); Martin Campbell-Kelly, William Aspray, and Nathan Ensmenger, *Computer: A History of the Information Machine*, 3rd ed. (Boulder, CO: Westview Press, 2014).

8. Steven Levy, *Hackers: Heroes of the Computer Revolution*, 25th anniversary ed. (Sebastopol, CA: O'Reilly, 2010); Paul Freiberger and Michael Swaine, *Fire in the Valley: The Making of the Personal Computer*, 2nd ed. (New York: McGraw-Hill, 2000); John Markoff, *What the Dormouse Said: How the Sixties Counterculture Shaped the Personal Computer Industry* (New York: Viking, 2005).

9. Theodore Roszak, *The Making of a Counter Culture: Reflections on the Technocratic Society and Its Youthful Opposition* (Garden City, NY: Anchor, 1969).

10. Mar Hicks, *Programmed Inequality: How Britain Discarded Women Technologists and Lost Its Edge in Computing* (Cambridge, MA: MIT Press, 2017); Tara McPherson, "U.S. Operating Systems at Mid-Century: The Intertwining of Race and UNIX," in *Race After the Internet*, ed. Lisa Nakamura and Peter A. Chow-White (New York: Routledge, 2012), 21–37; Joy Lisi Rankin, *A People's History of Computing in the United States* (Cambridge, MA: Harvard University Press, 2018).

11. Lauren Marshall Bowen, "The Limits of Hacking Composition Pedagogy," *Computers and Composition* 43 (2017): 1–14; Lilly Irani, "Hackathons and the Making of Entrepreneurial Citizenship," *Science, Technology, & Human Values* 40, no. 5 (2015): 799–824.

12. Ruha Benjamin, *Race after Technology: Abolitionist Tools for the New Jim Code* (Medford, MA: Polity, 2019), 28–29.

13. Claudia Lopez, "Speech Recognition Tech Is Yet Another Example of Bias," *Scientific American*, July 5, 2020, https://www.scientificamerican.com/article/speech-recognition-tech-is-yet-another-example-of-bias/.

14. Bruno Latour, *Reassembling the Social: An Introduction to Actor-Network-Theory* (New York: Oxford University Press, 2005), 80–81.

15. Natasha N. Jones, "Narrative Inquiry in Human-Centered Design: Examining Silence and Voice to Promote Social Justice in Design Scenarios," *Journal of Technical Writing and Communication* 46, no. 4 (2016): 471–92.

16. Lucy Suchman, "Working Relations of Technology Production and Use," *Computer Supported Cooperative Work* 2, nos. 1–2 (1993): 21–39.

17. Meredith Broussard, *Artificial Intelligence: How Computers Misunderstand the World* (Cambridge, MA: MIT Press, 2018); David Golumbia, *The Cultural Logic of Computation* (Cambridge, MA: Harvard, 2009).

18. Wendy Hui Kyong Chun, *Programmed Visions: Software and Memory* (Cambridge, MA: MIT Press, 2011); Lori Emerson, *Reading Writing Interfaces: From the Digital to the Bookbound* (Minneapolis: University of Minnesota Press, 2014); Alexander R. Galloway, *The Interface Effect* (Medford, MA: Polity, 2012); N. Katherine Hayles, *My Mother Was a Computer: Digital Subjects and Literacy Texts* (Chicago: University of Chicago Press, 2010); Lev Manovich, *Software Takes Command* (New York: Bloomsbury, 2013); Noah Wardrip-Fruin, *Express Processing: Digital Fictions, Computer Games, and Software Studies* (Cambridge, MA: MIT Press, 2009).

19. Robert R. Johnson, *User-Centered Technology: A Rhetorical Theory for Computers and Other Mundane Artifacts* (Albany: State University of New York Press, 1998).

20. Matthew Fuller and Andrew Goffey, *Evil Media* (Cambridge, MA: MIT Press, 2012), 11–12.

21. Matthew Fuller, *Beyond the Blip: Essays on the Culture of Software* (Brooklyn, NY: Autonomedia, 2003), 137–65.

22. Microsoft, "Microsoft Productivity Score," *Microsoft Docs*, June 3, 2021, https://docs.microsoft.com/en-us/microsoft-365/admin/productivity/content-collaboration?view=o365-worldwide.

23. Friedrich A. Kittler, "There Is No Software," in *Literature, Media, Information Systems*, ed. John Johnston (Amsterdam: Overseas Publishers Association, 1997), 151.

24. Kittler's description of modern computing architecture is very similar to that found in computer science textbooks. See, for example, Andrew Tanenbaum, *Structured Computer Organization* (Under Saddle River, NJ: Pearson, 2006).

25. Nick Montfort, "Continuous Paper: The Early Materiality and Workings of Electronic Literature," January 2005, https://nickm.com/writing/essays/continuous_paper_mla.html.

26. N. Katherine Hayles, "Print Is Flat, Code Is Deep: The Importance of Media-Specific Analysis," *Poetics Today* 25, no. 1 (2004): 67–90; Matthew G. Kirschenbaum, *Mechanisms: New Media and the Forensic Imagination* (Cambridge, MA: MIT Press, 2008).

27. Mark C. Marino, "Critical Code Studies," *electronic book review*, December 4, 2006, https://electronicbookreview.com/essay/critical-code-studies/; Kevin Brock, *Rhetorical Code Studies: Discovering Arguments in and around Code* (Ann Arbor: University of Michigan Press, 2019).

28. Michael L. Black, "Using Topic Modeling to Trace Sociocultural Influences on Software Development." *Digital Humanities Quarterly* 9, no. 3 (2015): http://www.digitalhumanities.org/dhq/vol/9/3/000224/000224.html.

29. Matthew G. Kirschenbaum, *Track Changes: A Literary History of Word Processing* (Cambridge, MA: Harvard University Press, 2018).

30. Deborah Brandt, *The Rise of Writing: Redefining Mass Literacy* (Cambridge: Cambridge University Press, 2015), chap. 1.

31. Safiya Umoja Noble, *Algorithms of Oppression: How Search Engines Reinforce Racism* (New York: New York University Press, 2018); Stuart A. Selber, *Multiliteracies for a Digital Age* (Carbondale: Southern Illinois University Press, 2004); Cynthia L. Selfe, *Technology and Literacy in the Twenty-First Century: The Importance of Paying Attention* (Carbondale: Southern Illinois University Press, 1999); Siva Vaidhyanathan, *The Googlization of Everything (And Why We Should Worry)* (Berkeley: University of California Press, 2011); Annette Vee, *Coding Literacy: How Computer Programming is Changing Writing* (Cambridge, MA: MIT Press, 2017).

32. Noble, *Algorithms of Oppression*, chap 2.

33. Emerson, *Reading Writing Interfaces*, introduction.

34. Michelle Bojar, "'Without human connection, school is a shadow,'" *Boston Globe*, December 10, 2020,https://www.bostonglobe.com/2020/12/10/metro/without-human-connection-school-is-shadow/; Matthew Finnegan, "Working From Home? Slow Broadband, Remote Security Remain Top Issues," *ComputerWorld*, October 5, 2020, https://www.computerworld.com/article/3584454/working-from-home-slow-broadband-remote-security-remain-top-issues.html.

35. Susan Leigh Starr, "Ethnography of Infrastructure," *American Behavioral Scientist* 43, no. 3 (1999): 384–85.

36. Bruno Latour, *We Have Never Been Modern*, trans. Catherine Porter (Cambridge, MA: Harvard University Press, 1993).

37. Theodor H. Nelson, *Computer Lib: You Can and Must Understand Computers Now* (Chicago: Hugo's Book Service, 1974).

38. Ceruzzi, *A History of Modern Computing*, 213.

39. Martin Campbell-Kelly, *From Airline Reservations to Sonic the Hedgehog: A History of the Software Industry* (Cambridge, MA: MIT Press, 2003), chap. 7.

40. Jeffrey Bardzell and Shaowen Bardzell, *Humanistic HCI* (San Rafael, CA: Morgan and Claypool, 2015).

41. Vee, *Coding Literacy*; Selfe, *Technology and Literacy in the Twenty-First Century*; Selber, *Multiliteracies for a Digital Age*.

42. Antonio Byrd, "'Like Coming Home': African Americans Tinkering and Playing toward a Computer Code Bootcamp," *College Composition and Communication* 71, no. 3 (2020): 426–52.

Chapter 1. On the Origins of User-Friendliness

1. Jeff Rubin and Dana Chisnell, *Handbook of Usability Testing*, 2nd ed. (Indianapolis, IN: Wiley, 2008), 4–6.

2. Steve Krug, *Don't Make Me Think! A Common Sense Approach to Web Usability*, 2nd ed. (Berkeley, CA: New Riders, 2006), chap 1.

3. The titles of his books alone reflect his ardent commitment to the principles of transparent design: *The Psychology of Everyday Things* (New York: Basic Books, 1988), *The Invisible Computer: Why Good Products Can Fail, the Personal Computer Is So Complex, and Information Appliances Are the Solution* (Cambridge, MA: MIT Press, 1998) and *Living with Complexity* (Cambridge, MA: MIT Press, 2010).

4. Norman, *The Psychology of Everything Things*, 183.

5. Steve Woolgar, "Technology as Cultural Artefact," in *Information and Communication Technologies: Visions and Realities*, ed. William H. Dutton (New York: Oxford University Press, 1996), 87–102.

6. Sam Edwards, "Why Is Software So Hard to Use?," *BYTE*, December 1983, 132.

7. Edwards, "Why Is Software So Hard to Use?"

8. Jay David Bolter and Richard Grusin, *Remediation: Understanding New Media* (Cambridge: MIT Press, 1999); Lev Manovich, *The Language of New Media* (Cambridge, MA: MIT Press, 2001); Jay David Bolter and Rachel Gromala, *Windows and Mirrors: Interaction Design, Digital Art, and the Myth of Transparency* (Cambridge, MA: MIT Press, 2003); Alexander R. Galloway, *The Interface Effect* (New York: Polity Press, 2012).

9. Bolter and Gromala, *Windows and Mirrors*, 38.

10. Bolter and Gromala, *Windows and Mirrors*, 44.

11. Walter Isaacson, *Steve Jobs* (New York: Simon and Schuster, 2011).

12. David Canfield Smith, Charles Irby, and Ralph Kimball, "Designing the Star User Interface," *BYTE*, April 1982, 246.

13. Macintosh advertising brochure, December 1983, Archive.org, https://archive.org/details/MAC_128K_BROCHURE_1983.

14. Mar Hicks, *Programmed Inequality: How Britain Discarded Women Technologists and Lost Its Edge in the Computing Revolution* (Cambridge, MA: MIT Press, 2017), 3.

15. Vannevar Bush, "As We May Think," *Atlantic Monthly*, July 1945, 101–8; Elizbeth R. Petrick, "The Historiography of Huma-Computer Interaction," *IEEE Annals of the History of Computing* 42, no. 4 (2020): 13. Both Wendy Chun (*Programmed Visions: Software and Memory* [Cambridge: MIT Press, 2011]) and Lori Emerson (*Reading Writing Interfaces: From the Digital to the Bookbound* [Minneapolis: University of Minnesota Press, 2014]) observe that Bush's arguments about the possibility of designing machine databases that complement and extend human reasoning resemble the rhetoric of more recent transparency advocates like Mark Weiser.

16. Emerson, *Reading Writing Interfaces*, chap. 2.

17. Thierry Bardini, *Bootstrapping: Douglas Engelbart, Coevolution, and the Origins of Personal Computing* (Stanford, CA: Stanford University Press, 2000), 209, 211n27.

18. Lauren Marshall Bowen, "The Limits of Hacking Composition Pedagogy," *Computers and Composition* 43 (2017): 1–14.

19. Saul Gorn, "Transparent-Mode Control Procedures for Data Communication, Using the American Standard Code for Information Interchange—a Tutorial," *Communications of the ACM* 8, no. 4 (1965): 203–6.

20. Harry J. Saal and Leonard J. Shustek, "Microprogrammed Implementation of Computer Measurement Techniques," in *Conference Record of the 5th Annual Workshop on Microprogramming*, ed. C. William Gear (New York: ACM, 1972), 42–50.

21. Edwin F. Hart, "User-Transparent Automatic Terminal Speed Selection," in *Proceedings of the 1974 Annual ACM Conference*, vol. 2, ed. Roger C. Brown (New York: ACM, 1974), 451–57.

22. Andrew S. Tanenbaum and Robbert Van Renesse, "Distributed Operating Systems," *Computing Surveys* 17 no. 4 (1984): 419–70.

23. For a more in-depth study of American computer magazine readers that describes similar trends in a different magazine, see also Laine Nooney, Kevin Driscoll, and Kera Allen, "From Programming to Products: *Softalk Magazine* and the Rise of the Personal Computer User," *Information & Culture* 55, no. 2 (2020): 105–29.

24. Carl Helmers, "On Entering Our Fourth Year," *BYTE*, September 1978, 6.

25. Chris Morgan, "The Hand-Held Computer," *BYTE*, January 1981, 8.

26. Lawrence J. Curran, "A Statement of Purpose," *BYTE*, July 1983. 4.

27. Carl Helmers, "Who Reads *BYTE*?," *BYTE*, October 1980, 8.

28. Stan Veit, who owned a computer store in New York City during the late 1970s and early '80s, wrote short profiles about his employees and customers. Although some were educated in the sciences, almost none had any prior experience with computers apart from basic classes they had taken in high school or college. See Stan Veit, *Stan Veit's History of the Personal Computer* (Asheville, NC: WorldComm, 1993), chap. 1.

29. Tri-Tek, Inc. "Get Your Bits Together," advertisement. *BYTE*, January 1976, 83.

30. Robert Nelson, "'Chip' Off the Olde PDP 8/E: The Intersil IM6100," *BYTE*, June 1976, 59.

31. J. M. Graetz, "The Origin of Spacewar," *Creative Computing*, August 1981, 60.

32. Dynabyte, "16384 BYTES for $485.00," advertisement, *BYTE*, May 1977, 119.

33. Stephen Wozniack, "The Apple-II", *BYTE*, May 1977, 39.

34. Wozniack, "The Apple-II," 40.

35. Dataquest, *Consolidated Data Base: U.S. Markets 1981–1990* (San Jose, CA: Dataquest Incorporated, 1986).

36. Frank Gens and Chris Christiansen, "Could 1,000,000 IBM PC Users Be Wrong?," *BYTE*, November 1983; Chris Rubin and Kevin Strehlo, "Why So Many Computers Look Like the 'IBM Standard,'" *Personal Computing*, March 1984.

37. Although they were marketed as "fully compatible," compatibly between IBM PCs and many clones was rarely as simple as "plug and play." See James Sumner, "Standards and Compatibility: The Rise of the PC Computing Platform," in *Whose Standards? Standardization, Stability, and Uniformity in the History of Information and Electrical Technologies*, ed. James Sumner and J. N. Gooday (London: Continuum, 2008), 101–27.

38. Kathryn S. Barley and James R. Driscoll, "A Survey of Data-Base Management Systems for Microcomputers," *BYTE*, November 1982, 219.

39. Wesper Micro Systems, "Wizard-spooler," advertisement, *BYTE*, November 1982, 10.

40. Abacus Data, Inc. "Informa X," advertisement, *BYTE*, November 1982, 248–49.

41. Netronics R&R Ltd., "Get IBM-PC Capacity at a Fraction of the IBM's Price," advertisement, *BYTE*, June 1983, 164.

42. Rixon, Inc., "The Rixon PC212A," advertisement, *BYTE*, May 1983, 431.

43. Chris Brown, "Top Brass Blow Taps to Hackers," *80 Microcomputing*, January 1982, 328.

44. Chris Rutkowski, "An Introduction to the Human Applications Standard Computer Interface," *BYTE*, October 1982, 302.

45. Rutkowski, "An Introduction to the Human Applications Standard Computer Interface," 299.

46. Lilly Irani, "Hackathons and the Making of Entrepreneurial Citizenship," *Science, Technology, & Human Values* 40, no. 5 (2015): 799–824.

47. Cynthia L. Selfe, *Technology and Literacy in the Twenty-First Century: The Importance of Paying Attention* (Carbondale: Southern Illinois University Press, 1999), 11.

48. Stuart A. Selber, *Multiliteracies for a Digital Age* (Carbondale: Southern Illinois University Press, 2004), 22–29.

49. Meredith Broussard, *Artificial Unintelligence: How Computers Misunderstand the World* (Cambridge, MA: MIT Press, 2018), chap 1.

50. Annette Vee, *Coding Literacy: How Computer Programming is Changing Writing* (Cambridge, MA: MIT Press, 2017), 3.

51. Vee, *Coding Literacy*, 27–9.

52. Siva Vaidhyanathan, *The Googlization of Everything (And Why We Should Worry)* (Berkeley: University of California Press, 2011), 109–11.

53. Terry Winograd and Fernando Flores, *Understanding Computers and Cognition: A New Foundation for Design* (Reading, MA: Addison-Wesley, 1986), 165.

54. Tara McPherson, "U.S. Operating Systems at Mid-Century: The Intertwining of Race and UNIX," in *Race After the Internet*, ed. Lisa Nakamura and Peter A. Chow-White (New York: Routledge, 2012), 21–37; Safiya Umoja Noble, *Algorithms of Oppression: How Search Engines Reinforce Racism* (New York: New York University Press, 2018).

55. Noble, *Algorithms of Oppression*, chap. 2.

56. "Control Your App Permissions on Android 6.0 & Up," Google Play Help, Google, accessed January 5, 2021, https://support.google.com/googleplay/answer/6270602?hl=en.

57. John Reardon, Alvaro Feal, Primal Wijesekera, Amit Elazari Bar On, Narseo Vallino-Rodriguez, and Serge Egelman, "50 Ways to Leak Your Data: An Exploration of Apps' Circumvention of the Android Permissions Systems," in *Proceedings of the 28th USENIX Conference on Security Symposium* (Santa Clara, CA: USENIX, 2019), 603–20.

58. Tom Warren, "Microsoft Will Now Release Major Windows 10 Updates Every March and September," *The Verge*, July 16, 2018, https://www.theverge.com/2017/4/20/15374864/microsoft-windows-10-update-september-2017.

59. Gordon Kelly, "Microsoft Warns Windows 10 Update Deletes Personal Data," *Forbes.com*, October 6, 2018, https://www.forbes.com/sites/gordonkelly/2018/10/06/microsoft-windows-10-update-lost-data-upgrade-windows-7-windows-xp-free

-upgrade/#2ea51c9c6fa4; Sarah Gray, "User Reports Problems with Microsoft Windows 10 Update That 'All But Destroyed' Laptop," *Fortune.com*, May 29, 2018, http://fortune.com/2018/05/29/microsoft-windows-10-april-2018-update-issues/.

60. Paul Lily, "Heads Up, Windows 10 Is Going to Claim 7GB of Storage for Future Updates," *PC Gamer*, January 10, 2019, https://www.pcgamer.com/heads-up-windows-10-is-going-to-claim-7gb-of-storage-for-future-updates/.

61. Sam Dean, "L.A. Is Suing IBM for Illegally Gathering and Selling User Data through Its Weather Channel App," *Los Angeles Times*, Jan 4, 2019, https://www.latimes.com/business/technology/la-fi-tn-city-attorney-weather-app-20190104-story.html.

62. Jordan McMahon, "Apple Had Way Better Options Than Slowing Down Your Phone," *Wired*, Dec 21, 2017, https://www.wired.com/story/apple-iphone-battery-slow-down/.

63. Emersion, *Reading Writing Interfaces*, chap. 1; Olga Goriunova and Alexei Shulgin, "Glitch," in *Software Studies: A Lexicon*, ed. Matthew Fuller (Cambridge, MA: MIT Press, 2008), 110–19.

64. Casey O'Donnell, "Mixed Messages: The Ambiguity of the MOD Chip and Pirate Cultural Production for the Nintendo DS," *New Media & Society* 16, no. 5 (2014): 737–52.

65. Ethan Gach, "Sony Is Still Trying to Stop People from Hacking the Vita, for Some Reason," *Kotaku*, August 27, 2019, https://kotaku.com/sony-is-still-trying-to-stop-people-from-hacking-the-vi-1837625652.

Chapter 2. The Sources of the Personal Computer Revolution

1. Nels Winkless III, "Personal Technology: More Strength in a Free Society," *Computer Notes*, August 1976, 7.

2. Steven B. Katz, "The Ethic of Expediency: Classical Rhetoric, Technology, and the Holocaust," *College English* 54, no. 3 (1992): 266.

3. Notably, Freiberger and Swaine's and Levy's books were originally published in 1984 and 1985, respectively. Both have at least three separate editions. That they continue to be reprinted in expanded editions is itself a testament to the durability of the narratives they provide. See Paul Freiberger and Michael Swaine, *Fire in the Valley: The Making of the Personal Computer* (New York: McGraw-Hill, 2000), Steven Levy, *Hackers: Heroes of the Computer Revolution* (Sebastopol, CA: O'Reilly, 2010), and John Markoff, *What the Dormouse Said: How the Sixties Counterculture Shaped the Personal Computer Industry* (New York: Penguin, 2004).

4. Brian Pfaffenberger, "The Social Meaning of the Personal Computer Revolution: Or, Why the Personal Computer Revolution Was No Revolution," *Anthropological Quarterly* 61, no. 1 (1988): 43.

5. Elizabeth R. Petrick, "Imagining the Personal Computer: Conceptualizations of the Homebrew Computer Club 1975–1977," *IEEE Annals of Computer History* 39, no. 4 (2017): 27–39.

6. Joy Lisi Rankin, *A People's History of Computing in the United States* (Cambridge, MA: Harvard University Press, 2018), 228–29.

7. Rankin, *A People's History of Computing in the United States*, 237–38.

8. Luke Stark, "Here Come the 'Computer People': Anthropomorphosis, Command, and Control in Early Personal Computing," *IEEE Annals of the History of Computing* 42, no. 4 (2020): 53–70.

9. Susan Leigh Star, "Ethnography of Infrastructure," *American Behavioral Scientist* 43, no. 3 (1999): 384–85.

10. Levy, *Hackers*, 26–38.

11. Levy, *Hackers*, 27.

12. Levy, *Hackers*, ix,

13. Levy, *Hackers*, 29.

14. Levy, *Hackers*, 136.

15. Burton Grad, "A Personal Recollection: IBM's Unbundling of Software and Services," *IEEE Annals of the History of Computing* 24, no. 1 (2002): 64–71.

16. Levy, *Hackers*, 30.

17. Katharine Davis Fishman, *The Computer Establishment* (New York: McGraw-Hill, 1981), 37.

18. Fishman, *The Computer Establishment*, 44.

19. Katz, "The Ethic of Expediency," 263.

20. Levy, *Hackers*, 125–27.

21. Fred Turner, *From Counterculture to Cyberculture: Stewart Brand, the Whole Earth Network, and the Rise of Digital Utopianism* (Chicago: University of Chicago Press, 2006), 135–36.

22. Stewart Brand, "Spacewar: Fantastic Life and Symbolic Death among the Computer Bums," *Rolling Stone*, December 7, 1972, 50–58.

23. Turner, *From Counterculture to Cyberculture*, 89–91.

24. *Whole Earth Catalog*, fall 1968, 2.

25. Turner, *From Counterculture to Cyberculture*, 83–84

26. Turner, *From Counterculture to Cyberculture*, 92–97.

27. Turner, *From Counterculture to Cyberculture*, 97–98.

28. Turner, *From Counterculture to Cyberculture*, 119.

29. Turner, *From Counterculture to Cyberculture*, chap 7.

30. Ester Dyson, George Gilder, George Keyworth, and Alvin Toffler, "Cyberspace and the American Dream: A Magna Carta for the Knowledge Age," August 1994, http://www.pff.org/issues-pubs/futureinsights/fi1.2magnacarta.html.

31. Peter H. Lewis, "Cyberspace Prophets Discuss Their 'Revolution' Face to Face," *New York Times*, August 23, 1995, 17.

32. Dyson, Gilder, Keyworth, and Toffler, "Cyberspace and the American Dream," 172.

33. Stuart A. Selber, *Multiliteracies for a Digital Age* (Carbondale: Southern Illinois University Press, 2004), 24–25.

34. Theodor H. Nelson, *Computer Lib / Dream Machines* (Chicago: Hugo's Book Service, 1974), 2.

35. Nelson, *Computer Lib / Dream Machines*, 2–3.

36. Nelson, *Computer Lib / Dream Machines*, front cover.

37. Nelson, *Computer Lib / Dream Machines*, 54.

38. Nelson, *Computer Lib / Dream Machines*, 53.

39. Nelson, *Computer Lib / Dream Machines*, 2.

40. Nelson, *Computer Lib / Dream Machines*, 7.

41. Levy, *Hackers*, 179–82; Markoff, *What the Dormouse Said*, 196.

42. Donald A. Norman, "Cognitive Engineers," *User Centered System Design*, ed. Donald A. Norman and Stephen W. Draper (Hillsdale, NJ: Lawrence Erlbaum Associations, 1986), 49.

43. Ivan Illich, *Tools for Conviviality* (New York: Harper and Row, 1973), 11

44. Illich, *Tools for Conviviality*, 21.

45. Illich, *Tools for Conviviality*, 7–8.

46. Illich, *Tools for Conviviality*, 36–9.

47. Illich, *Tools for Conviviality*, 33.

48. Illich, *Tools for Conviviality*, 3–7.

49. Illich, *Tools for Conviviality*, 13.

50. Illich, *Tools for Conviviality*, 22.

51. Illich, *Tools for Conviviality*, 25–26.

52. Richard Barbrook and Andy Cameron, "The Californian Ideology," *Science as Culture* 6, no.1 (1996): 44–72.

53. Illich, *Tools for Conviviality*, 86.

54. Levy, *Hackers*, 170.

55. Turner, *From Counterculture to Cyberculture*, 114.

56. LeRoy Finkle, "IBM Wants You," *People's Computer Company*, July 1975, 8.

57. *People's Computer Company*, May 1973, 13.

58. "Educational Computer Buyers Guide," *People's Computer Company*, October 1972, 13–15; "Writing Bid Specs," pt. 1, *People's Computer Company*, February 1983, 12; LeRoy Finkel, "Writing Bid Specs," pt. 2, April 1973, 8–9; Bob Albrecht, "How to Buy an Edu System," *People's Computer Company*, April 1973, 21.

59. Gregory Yob, "Buy Your TV Set Something Nice, Like a Computer, or Buy a Brain For Your Boob Tube," *People's Computer Company*, September 1973, 5.

60. Lee Felsenstein, "Convivial Design," *People's Computer Company*, December 1974, 15.

61. *People's Computer Company*, January 1975, 2.

62. Levy, *Hackers*, 198–99; Markoff, *What the Dormouse Said*, 273–75.

63. Petrick, "Imagining the Personal Computer," 29.

64. Levy, *Hackers*, 197–98; Markoff, *What the Dormouse Said*, 31–39, 196; Turner, *From Counterculture to Cyberculture*, 102.

65. Fred More, "'It's a Hobby,'" *Homebrew Computer Club Newsletter*, June 7, 1976, 1.

66. Fred Moore, "The Club Is All of Us," *Homebrew Computer Club Newsletter*, April 12, 1975, 2.

67. Felsenstein's, Draper's, Pittman's, and Wozniak's information appear in the April 12, 1975, issue, Warren's in the May 10, 1975, issue, and Lawson's in the July 5, 1975, issue.

68. Lee Felsenstein, "Excerpts from a Chalk Talk on the Tom Swift Terminal," *Homebrew Computer Club Newsletter*, July 5, 1975, 2.

69. Jonathan Titus, "Build the Mark 8 Computer," *Radio Electronics*, July 1974, 30–32.

70. In these next few paragraphs, I refer to scans of the newsletter compiled by Bryan K. Blackburn and available at "Mark-8 Documentation," *Bryan's Old Computers*, accessed March 25, 2018, http://bytecollector.com/m8_docs.htm.

71. Dennis Allison, "Build Your Own BASIC," *People's Computer Company*, March 1975, 6–7.

72. "Build Your Own BASIC—Revived," *People's Computer Company*, July 1975, 12; Dennis Allison, "Design Notes for TINY BASIC," *People's Computer Company*, September 1975, 15–18.

73. Pitt advertised his port of Allison's Tiny BASIC interpreter for the Motorola 6800 for $5. See "Itty Bitty Computers Tiny BASIC for the 6800," *BYTE*, July 1976, 76.

74. "Software Contest," *Computer Notes*, April 1975, 4.

75. Bill Gates, "July Software Content Winners Announced," *Computer Notes*, August 1975, 7.

76. "Altair BASIC—Up and Running," *Computer Notes*, April 1975, 1.

77. "Software Prices," *Computer Notes*, April 1975, 3.

78. "New Products," *Computer Notes*, April 1975, 4.

79. H. Edward Roberts, "Letter from the President," *Computer Notes*, October 1975, 3–4.

80. "Software Prices," *Computer Notes*, October 1975, 15.

81. Barbara Sims, "Altair Service Dept." *Computer Notes*, August 1975, 3.

82. Bill Gates, "An Open Letter to Hobbyists," *Computer Notes*, February 1976, 3.

83. Kevin Driscoll, "Professional Work for Nothing: Software Commercialization and 'An Open Letter to Hobbyists,'" *Information & Culture* 50, no. 2 (2015): 257–83.

84. Driscoll, "Professional Work for Nothing: Software Commercialization and 'An Open Letter to Hobbyists,'" 265–67.

85. Bill Gates, "A Second and Final Letter," *Computer Notes*, April 1976, 5.

86. Sarah A. Peterson, "Electronics Spawns a New Breed of Tycoon," *U.S. News and World Report*, December 27, 1982, 70–72.

87. Freiberger and Swaine, *Fire in the Valley*, 228.

88. "History," *Trenton Computer Festival*, accessed March 25, 2020, https://tcf-nj.org/history/.

89. Brian Bagnall, *Commodore: A Company on the Edge*, (Winnipeg: Variant Press, 2011), 125.

90. Levy, *Hackers*, 272. In less provocative language, Freiberger and Swaine similarly characterize the event as attended mostly by hobbyists (*Fire in the Valley*, 330).

91. David H. Ahl, "The First West Coast Computer Faire," *Creative Computing*, July/August 1977, 22–27.

92. Ahl, "The First West Coast Computer Faire," 24.

93. Theodore H. Nelson, "Those Unforgettable Next Two Years," *The Proceedings of the First Annual West Coast Computer Faire*, ed. Jim C. Warren Jr. (Palo Alto, CA: Computer Faire Inc., 1977), 22.

94. Nelson, "Those Unforgettable Next Two Years," 25.

95. Nelson, "Those Unforgettable Next Two Years," 25.

96. Annette Milford, "Computer Power of the Future," *Computer Notes*, April 1976, 7.

97. "The Mothership Apple Advertising and Brochure Gallery, 1976–1979," *The Mothership!*, accessed March 26, 2018, http://www.macmothership.com/gallery/gallery1.html.

98. *Apple II Reference Manual* (Cupertino, CA: Apple Computer, Inc., 1978); Jef Raskin, *Apple II BASIC Programming Manual* (Cupertino, CA: Apple Computer, Inc., 1978).

99. "Apple's New Software," *Intelligent Machines Journal*, June 11, 1979, 22; *Apple Software Bank: Contributed Programs*, vols. 1 and2 (Cupertino, CA: Apple Computer, Inc., 1978), 1.

100. Owen W. Linzmayer, *Apple Confidential 2.0: The Definitive History of the World's Most Colorful Company* (San Francisco: No Starch Press, 2004), 14–15; Rankin, *A People's History of Computing in the United States*, 237–38.

101. Apple Computer, Inc., "Introducing Apple II," advertisement, *BYTE*, June 1979, 14–15.

102. Apple Computer, Inc., "Working Its Way through Colleges," advertisement. *Creative Computing*, May 1980, inside front cover; Apple Computer, Inc., "Reddy Chirra Improves His Vision with an Apple," advertisement. *BYTE*, July 1981, 12–13; Apple Computer, Inc., "Not All Business Graphics Are Alike," advertisement. *BYTE*, November 1982, 264–65.

103. John Solomon, "The Remote Work Revolution—and How Chromebooks and Chrome OS help," Google Cloud Blog, June 16, 2020, https://cloud.google.com/blog/products/chrome-enterprise/the-remote-work-revolution.

104. "Meltdown and Spectre," https://meltdownattack.com; Nathaniel Mott, "Microsoft Confirms 'PrintNightmare' Vulnerability Affects All Windows Versions," *PC Magazine*, July 2, 2021, https://www.pcmag.com/news/microsoft-confirms-printnightmare-vulnerability-affects-all-windows-versions.

105. Lilly Irani, "Hackathons and the Making of Entrepreneurial Citizenship," *Science Technology, & Human Values* 49, no. 5 (2015): 814–15.

106. "Free Software Is a Matter of Liberty, Not Price," Free Software Foundation, accessed January 9, 2021, https://www.fsf.org/about/.

107. Coraline Ada Ehmke, "The New Normal: Codes of Conduct in 2015 and Beyond," *Model View Culture*, December 16, 2015, https://modelviewculture.com/pieces/the-new-normal-codes-of-conduct-in-2015-and-beyond.

108. Daniel Oberhaus, "The Culture War Comes to Linux," *VICE*, September 26, 2018, https://www.vice.com/en/article/yw43kj/what-happens-if-linux-developers-remove-their-code; Eric S. Raymond, "Why Hackers Must Reject the SJWs," Armed and Dangerous, November 13, 2015, http://esr.ibiblio.org/?p=6918.

Chapter 3. Appliance Computing

1. Wayne Green, "Where We're Coming from/Where We're Bound," *80 Microcomputing*, October 1980, 9.

2. Takashi Hikino, Andrew Von Nordenflycht, and Alfred D. Chandler Jr., *Inventing the Electronic Century: The Epic Story of the Consumer Electronics and Computer Industries* (Cambridge, MA: Harvard University Press, 2005), 135.

3. Early editions of Paul Freiberger and Michael Swaine's *Fire in the Valley* discuss the TRS-80 in a chapter titled "The McDonald's of Computing." The latest edition, however, is slightly more complimentary of the machine and moves and reorganizes material from that chapter into a new one called "Retailing the Revolution." Even the more complimentary version is still full of backhanded remarks such as that it helped to "expand the market" despite its "toy like" design. For comparison, see Paul Freiberger and Michael Swaine, *Fire in the Valley: The Making of the Personal Computing*, 2nd ed. (New York: McGraw-Hill, 2000), 239–49, and Michael Swaine and Paul Freiberger, *Fire in the Valley: The Birth and Death of the Personal Computer*, 3rd ed. (Dallas: The Pragmatic Bookshelf, 2014), 210–14.

4. Paul E. Ceruzzi, *A History of Modern Computing*, 2nd ed. (Cambridge, MA: MIT Press, 2003): 264; Martin-Campbell Kelly, William Aspray, and Nathan Ensmenger, *Computer: History of the Information Machine*, 3rd ed. (Boulder, CO: Westview Press, 2014), 241.

5. Lori Emerson, *Reading Writing Interfaces: From the Digital to the Bookbound* (Minneapolis: University of Minnesota Press, 2014), chap. 2.

6. Mark Weiser, "The Computer for the 21st Century," *Scientific American*, September 1991, 94.

7. Paul Dourish and Genevieve Bell, *Divining a Digital Future: Mess and Mythology in Ubiquitous Computing* (Cambridge, MA: MIT Press, 2011), 23–39.

8. Carl Helmers, "Reflections on Entry into Our Third Year," *BYTE*, September 1977, 6.

9. For examples of praise of appstores as a platform improving user choice, see Pelle Snickars, "A Walled Garden Turned into a Rain Forest," *Moving Data: The iPhone and the Future of Media*, ed. Pelle Snickars and Patrick Vonderau (New York: Columbia University Press, 2012), 155–68, and Barbara Flueckiger, "The iPhone Apps: A Digital Culture of Interactivity," *Moving Data: The iPhone and the Future of Media*, ed. Pelle Snickars and Patrick Vonderau (New York: Columbia University Press, 2012), 171–83.

10. Nick Montfort and Ian Bogost, *Racing the Beam: The Atari Video Computer System* (Cambridge, MA: MIT Press, 2009), vii.

11. Tarleton Gillespie, "The Politics of Platforms," *New Media & Society* 12, no. 3 (2010): 347–64.

12. Steven Levy, *Hackers: Heroes of the Computer Revolution* (Sebastopol, CA: O'Reilly, 2010).

13. Thomas Haigh, "Remembering the Office of the Future: The Origins of Word Processing and Office Automation," *IEEE Annals of the History of Computing* 28, no. 4 (2006): 19.

14. For some extended examples, see "Soon: A Computer in Every Home?" *U.S. News & World Report*, November 21, 1977, "Thinking Small: Little Whizzes Raise the Specter of Buggy Whips," *Time*, February 20, 1978, Christopher Byron, "Now the Office of Tomorrow: Technology's Dazzling Breakthroughs Shake up the White-Collar World," *Time*, November 17, 1980, Andrew Pollack, "Next, a Computer on Every Desk," *New York Times*, August 23, 1981, and William D. Marbach, "Invasion of the Computers," *Newsweek*, December 28, 1981.

15. Thomas O'Toole, "The Smart Machine Revolution: Miniaturization Brings Computers to the Home," *Washington Post*, June 5, 1977.

16. "Coming: Another Revolution in Use of Computers," *U.S. News & World Report*, July 19, 1976, 54.

17. In addition to the *Time* story I have quoted, see also William D. Marbach, "To Each His Own Computer," *Newsweek*, February 22, 1982, and Judith B. Gardner, "Computers for the Masses," *U.S. News & World Report*, December 27, 1982.

18. Jay Cocks, "The Computer Moves In," *Time*, January 3, 1983, 14.

19. Cocks, "The Computer Moves In," 22.

20. Schuyten makes the same comment over a year later, suggesting his impression of the market had changed little in the interim. See Peter J. Schuyten, "Home Computer: Demand Lags," *New York Times*, June 7, 1979, D2, and Peter J. Schuyten, "Home-Computer Software Lags" *New York Times*, November 26, 1980, D1.

21. Peter J. Schuyten, "A Computer to Call Your Own," *New York Times*, June 23, 1980; Peter J. Schuyten, "Bringing Your System Home," *New York Times*, June 30, 1980.

22. See for example, Anne Fisher Ruth, "Does a Home Computer Make Sense for You?," *McCall's*, September 1982, and Herbert Kohl, "Should You Buy Your Child a Computer?," *Saturday Evening Post*, December 1982.

23. In addition to those sources cited in the remainder of this section, for a good overview of how journalists understood the split between appliance and general purpose computers in the market see David H. Ahl, "The First Decade of Personal Computing," *Creative Computing*, November 1984, 40–34.

24. Michael Swaine, "The Computer Store—Year Zero," *InfoWorld*, June 8, 1981.

25. Patty Rust, "Mystery Shopper," *Creative Computing*, June 1979.

26. Anonymous, "Disillusioned Computer-Store Employee Recites Sad Saga," *InfoWorld*, July 26, 1982.

27. In addition to Schuyten, Rust, and others cited here, see also "A Bright New World of Computers," *New York Times*, June 4, 1981, C8, and Joe Desposito, "Which One Is For You?," *Popular Electronics*, May 1982.

28. Betsy Staples and John J. Anderson, "Undercover Consumer," *Creative Computing*, June 1984.

29. Prices quoted from advertisements featured in *BYTE* magazine. The Apple II price appears in the June1977 issue. The TRS-80 prices appear in the October 1977 issue.

30. As I discussed in more detail in the text, Commodore did not initially advertise in American personal computing magazines. Price quoted from Brian Bagnall, *Commodore: A Company on the Edge* (Winnipeg: Variant Press), 114.

31. The Atari prices are quoted from an advertisement appearing in the January 1980 of *BYTE*. The Texas Instruments prices are quoted from Joseph Nocera, "Death of a Computer," pt. 1, *InfoWorld*, June 4, 1984, 62. As I note in the text, Texas Instruments did not advertise in computing magazines.

32. All prices quoted from advertisements in the July 1977 issue of *BYTE*.

33. The description of the TRS-80's design and its usage is summarized from Green, "Where We're Coming from/Where We're Bound," 8, Stan Veit, *Stan Veit's History of the Personal Computer* (Asheville, NC: WorldComm, 1993), 153–66, and Theresa M. Welsh and David Welsh, *Priming the Pump: How TRS-80 Enthusiasts Helped Spark the PC Revolution* (Ferndale, MI: Seeker Books, 2013).

34. The description of the PET's design and its usage in this paragraph is summarized from Veit, *Stan Veit's History of the Personal Computer*, 176, and Bagnall, *Commodore*, 65–94.

35. Commodore Business Machines, *User's Manual for CBM Dual Drive Floppys* (West Chester, PA: Commodore Business Machines, 1980), 3.

36. The description of the design and usage of the Atari 400 and 800 in this paragraph is summarized from Jamie Lendino, *Breakout: How Atari 8-Bit Computers Defined a Generation* (New York: Ziff Davis, 2017), 24–40, and Chris Morgan, "Atari's New Hybrid Computers," *onComputing*, Fall 1979.

37. The description of the design and usage of the TI-99/4 and TI-99/4A in this paragraph is summarized from Nocera, "Death of a Computer," pt. 1.

38. Chris Morgan, "The Texas Instruments 99/4 Personal Computer," *onComputing*, Fall 1980, 29.

39. Bruce Van Voorst, "Price War in Small Computers," *Time*, September 20, 1982.

40. Paul Freiberger, "Commodore Founder Tramiel: PETS for the World Market," *InfoWorld*, April 26, 1982.

41. Unless otherwise noted, the general history of the TRS-80 provided in this section is synthesized from David Ahl, "Tandy Radio Shack Enters the Magic World of Computers," *Creative Computing*, November 1984, Chris Brown, "The Tandy Story," *80 Microcomputing*, January 1980. Jonathan Erickson, "The Men Behind the TRS-80s," *Popular Computing*, December 1981, Paul Freiberger, "Leather to Micros: The Rise of Tandy," *InfoWorld*, August 31, 1981, Michael Swaine, "How the TRS-80 Was Born," *InfoWorld*, August 31, 1981, and Ron White, "The Tandy Story," *80 Microcomputing*, August 1987.

42. Irvin Farman, *Tandy's Money Machine: How Charles Tandy Built Radio Shack into the World's Largest Electronics Chain* (Chicago: Mobium Press, 1992), 405.

43. Nancy Robertson, "Lew Kornfeld," *80 Microcomputing*, March 1981, 87.

44. Bert Latamore, "Microcomputers—Business or Pleasure," *80 Microcomputing*, July 1981, 90.

45. David Needle, "Q&A: Steve Leininger," *InfoWorld*, July 16, 1984.

46. Ahl, "Tandy Radio Shack Enters the Magic World of Computers," 294.

47. Wes Thomas, "Radio Shack's $600 Home Computer," *Creative Computing*, September / October 1977, 94.

48. The list referenced here appears in the December 31, 1975 issue.

49. White, "The Tandy Story," 52.

50. Thomas, "Radio Shack's $600 Home Computer," 94.

51. Radio Shack, "The First Complete Low-Cost Microcomputer for Home, Business or Education!," advertisement. *BYTE*, October 1977, 43.

52. Radio Shack, "Radio Shack's Personal Computer *System*?," advertisement, *BYTE*, June 1978, 73.

53. Radio Shack, *TRS-80 Microcomputer Catalog* (Fort Worth, TX: Tandy Corporation, 1978), 3.

54. Tom Williams, "TRS-80: A Consumer's Computer?," *People's Computers*, March / April 1978.

55. Letters to the editor *People's Computers*, July/August 1978, 4–5.

56. Brown, "The Tandy Story," 30.

57. Dan Fylstra, "The Radio Shack TRS-80: An Owner's Report," *BYTE*, April 1978, 58–59.

58. Ed Juge, "The TRS-80: How Does it Stack up?," *Kilobaud Microcomputing*, January 1978, 45.

59. Juge, "The TRS-80," 49.

60. Juge, "The TRS-80," 49.

61. Thomas Dwyer and Margot Critchfield, "The TRS-80 Level II Computer: A Grown-Up Field Evaluation," *People's Computers*, September / October 1978, 7.

62. Wayne Green, "Painful Fact of Life," *80 Microcomputing*, August 1980, 8.

63. Jerry Pournelle, "Omikron TRS-80 Boards, NEWDOS+, and Sundry Other Matters," *BYTE*, July, 1980, 204.

64. Veit, *Stan Veit's History of the Personal Computer*, 170.

65. Unless otherwise noted, the general history of the PET provided in this section is synthesized from Bagnall, *Commodore*, Robert J. Crowell, "A Commodore Perspective," *Compute!*, fall 1979, and Freiberger, "Commodore Founder Tramiel."

66. Richard Simpson, "The KIM Forum," *Kilobaud Microcomputing*, July 1977; Gregg Williams and Rob Moore, "The Apple Story: Early History," *BYTE*, December 1984.

67. David Needle, "Strategy and a Preview of the Commodore 'Emulators,'" *InfoWorld*, April 26, 1982.

68. Veit, *Stan Veit's History of the Personal Computer*, 177; Ralph Wells, "PET's First Report Card: An Objective Evaluation," *Kilobaud Microcomputing*, May 1978; Needle, "Strategy and a Preview of the Commodore 'Emulators.'"

69. "The $595 PET," *People's Computers*, September / October 1977, 22.

70. Sheila Clarke, "A PET for Every Home: A Look at the Commodore PET 2001," *Kilobaud Microcomputing*, September 1977, 42.

71. "The $595 PET," 23.

72. "The $595 PET," 24.

73. Clarke, "A PET for Every Home," 42.

74. "The $595 PET," 26.

75. "The $595 PET," 26.

76. Wells, "PET's First Report Card," 29.

77. Wells, "PET's First Report Card," 29.

78. Dan Fylstra, "User's Report: The PET 2001," *BYTE*, March 1978, 126.

79. Peter Martin, "My Pet and I: Testing the Appliance of the Future, a Home Computer, a *Money* Editor Finds the Future Doesn't Always Work," *Money*, May 1978.

80. Elizabeth Peer, "How to Work the Thing," *Newsweek*, February 22, 1982.

81. Commodore Business Machines, "The Great American Solution Machine," advertisement. *Compute!*, February 1981, back cover; Commodore Business Machines, "VIC-20: The Friendly Computer," *BYTE*, January 1982, 75.

82. For descriptions and analysis of the rhetoric and imagery use in examples of early personal computing promotional material, see Luke Stark, "Here Come the 'Computer People': Anthropomorphosis, Command, and Control in Early Personal Computing," *IEEE Annals of the History of Computing* 42, no. 4 (2020): 53–70.

83. John Victor, "Atari Computers: The Ultimate Teaching Machines?," *Compute!*, fall 1979, 62.

84. Morgan, "Atari's New Hybrid Computers," 51.

85. Ted Nelson, "The Atari Machine," *Creative Computing*, June 1980, 34.

86. Frank J. Derfler Jr., "Moonshine, Dixie and the Atari 800," *Kilobaud Microcomputing*, September 1980.

87. Ken Skier, "The Atari 800 Personal Computer," *onComputing*, Summer 1980, 25.

88. Michael S. Tomczyk, "Atari's Marketing Vice President Profiles the Personal Computer Market," *Compute!*, July / August 1980, 17.

89. Tomczyk, "Atari's Marketing Vice President Profiles the Personal Computer Market," 16.

90. A column by this name briefly appeared in *Atari Connection* beginning with the first issue in spring 1981.

91. Scott Mace, "Atari Cognoscenti Gather and Gab," *InfoWorld*, July 26, 1982, 22.

92. Mace, "Atari Cognoscenti Gather and Gab," 23.

93. Mark J. Wood, *The Video Game Explosion: A History from PONG to Playstation and Beyond* (Westport, CT: Greenwood Press, 2008), 56.

94. Atari's ROM cartridges are discussed as a copy protection technology in Jake Commander and G. Michael Vose, "How Safe Is Your Software?" *Kilobaud Microcomputing*, July 1982, 61.

95. Nick Montfort and Ian Bogost, *Racing the Beam: The Atari Video Computer System*, (Cambridge, MA: MIT Press, 2009), chap. 6.

96. Robert C. Lock, editorial, *Compute!*, January 1982, 10; David D. Thornburg, "Computers and Society," *Compute!*, June 1982, 14.

97. Skier, "The Atari 800 Personal Computer," 59.

98. Nelson, "The Atari Machine," 60.

99. Thom Hogan, "The Atari 800," *InfoWorld*, May 11, 1981, 36.

100. Michael S. Tomczyk, *The Home Computer Wars: An Insider's Account of Commodore and Jack Tramiel* (Greensboro, NC: Compute! Publications, 1984), 110.

101. The general history of the TI-99/4A provided in this section is synthesized from Scott Mace, "TI Retires from Home-Computer Market," *InfoWorld*, November 21, 1983, David I . Ahl, "Texas Instruments," *Creative Computing*, March 1984, Nocera, "Death of a Computer", pt. 1, and Joseph Nocera, "Death of a Computer," pt. 2, *InfoWorld*, June 11, 1984.

102. "TI to Introduce Personal Computer at Summer CES," *Intelligent Machines Journal*, June 11, 1979.

103. The lower number was quoted by a market-research firm and the higher internally by Texas Instruments executives. See, respectively Scott Mace, "Texas Instruments in the Saddle," *InfoWorld*, May 30, 1983, 26, Nocera, "Death of a Computer," pt. 2, 64.

104. Mace, "TI Retires from Home-Computer Market," 22.

105. Texas Instruments, "Buy now and get FREE Solid State Software Libraries," advertisement, *BYTE*, September 1980, 49.

106. Texas Instruments, "Join TI as Its headquarters for Consumer Electronic Products and Enjoy a Bold New World," advertisement, *BYTE*, July 1980, 239.

107. Texas Instruments, "Announcing Texas Instruments Author Incentive Program," advertisement, *BYTE*, September 1980, 105.

108. Texas Instruments, "TI's Home Computer. This is the one," advertisement, *BYTE*, September 1982, 103.

109. Nocera, "Death of a Computer," pt. 1, 62.

110. Nocera, "Death of a Computer," pt. 1, 62.

111. Nocera, "Death of a Computer," pt. 1, 60.

112. Nocera, "Death of a Computer," pt. 1, 61.

113. Morgan, "The Texas Instruments 99/4 Personal Computer"; Fred Gray, "The TI 99/4A," *Creative Computing*, May 1983.

114. Veit, *Stan Veit's History of the Personal Computer*, 198–99.

115. Richard Milewski, "TI Ignores Third-Party Software," *InfoWorld*, July 21, 1980, 28.

116. Scott Mace, "Texas Instrument's GROM Produces 'Solid-State Software,'" *InfoWorld*, May 30, 1983, 29.

117. Nocera, "Death of a Computer," pt. 2, 63–64.

118. Van Voorst, "Price War in Small Computers."

119. Peggy Zientara, "When Was the Last Time You Used Your Micro?" *InfoWorld*, November 28, 1983, 36.

120. Bro Uttal, "Sudden Shake-Up in Home Computers," *Fortune*, July 11, 1983.

121. Andrew Pollack, "The Coming Crisis in Home Computers," *New York Times*, June 19, 1983.

122. Emerson, *Reading Writing Interfaces*, chap 1.

123. For examples of software being removed or radically changed without warning, see Cyrus Farviar, "DRM Be Damned: How to Protect Your Amazon E-Books from Being Deleted," *Ars Technica*, October 25, 2012, https://arstechnica.com/gadgets/2012/10/drm-be-damned-how-to-protect-your-amazon-e-books-from-being-deleted/, and Kyle Orland, "Apple Forces Nude Immigrants to Cover up in iPad version of Papers, Please," *Ars Technica*, December 12, 2014, https://arstechnica.com/gaming/2014/12/apple-forces-nude-immigrants-to-cover-up-in-ipad-version-of-papers-please/.

124. Tony Room and Rachel Lerman, "Amazon Suspends Parler, after Google and Apple Also Take Action," *Washington Post*, January 11, 2021, https://www.washingtonpost.com/technology/2021/01/09/amazon-parler-suspension/.

125. David D. Thornburg and Betty J. Burr, "Computers and Society," *Compute!*, November / December 1980, 12–13.

Chapter 4. IBM, Apple, and a Computer Literacy Crisis

1. Annette Vee, *Coding Literacy: How Computer Programming Is Changing Writing* (Cambridge, MA: MIT Press, 2017), 27–34.

2. Andrea diSessa, *Changing Minds: Computers, Learning, and Literacy* (Cambridge, MA: MIT Press, 2000), 4–5, 113–14.

3. Cynthia L. Selfe, *Technology and Literacy in the Twenty-First Century: The Importance of Paying Attention* (Carbondale: Southern Illinois Press, 1999); Stuart A. Selber, *Multiliteracies for a Digital Age* (Carbondale: Southern Illinois Press, 2004).

4. John Trimbur, “Literacy and the Discourse of Crisis,” *The Politics of Writing Instruction: Postsecondary*, ed. Richard Bullock and John Trimbur (Portsmouth, NH: Boynton / Cook, 1991), 281.

5. Thom Hogan, “From Backyard to Big Time,” *Creative Computing*, November 1983.

6. David D. Thornburg, “Computers and Society,” *Compute!*, June 1981, 10.

7. Robert Cowen, “Cottage Computing: Why Glorify the Trivial?,” *Technology Review*, November / December 1981, 20.

8. Robert Cowen, “Home Computing Revisited,” *Technology Review*, May / June 1982, 7.

9. Andrew R. Molnar, “The Next Great Crisis in American Education: Computer Literacy,” *Journal of Education Technology Systems* 7, no. 3 (1979): 280–83.

10. Marc Levinson, “Computers: Dealing with Terminal Phobia,” *Time*, July 19, 1982.

11. Craig Brod, *Technostress: The Human Cost of the Computer Revolution* (Reading, MA: Addison-Wesley, 1984).

12. Henry M. Levin, “High-Tech Requires Few Brains,” *Washington Post*, January 30, 1983.

13. Ed Martino, “Unorthodox Approach to Understanding Computers: Don’t Start With Programming!,” *InfoWorld*, November 10, 1980, 10.

14. Martin Campbell-Kelly, *From Airline Reservations to Sonic the Hedgehog: A History of the Software Industry* (Cambridge, MA: MIT Press, 2003), chap. 7.

15. Martino, “Unorthodox Approach to Understanding Computers,” 11.

16. Lee The, “Squaring off over Computer Literacy,” *Personal Computing*, September 1982, 69.

17. The, “Squaring off over Computer Literacy,” 63.

18. Jim Edlin, “The Mass Market Micro,” *InfoWorld*, December 8, 1980, 14.

19. Edlin, “The Mass Market Micro,” 26.

20. Edlin, “The Mass Market Micro,” 27.

21. Jim Edlin, “The Demise of Documentation,” *InfoWorld*, February 2, 1981; Jim Edlin, “Micros Should ‘Sell Themselves,’ ” *InfoWorld*, March 30, 1981.

22. Dorothy Heller, “Mirror, Mirror, on the Wall, who’s the ‘User-Friendliest’ of Them All?,” *InfoWorld*, April 5, 1982, 11.

23. For an in-depth discussion of how these differences could affect even a simple program, see Nick Monfort, Patsy Baudoin, John Bell, Ian Bogost, Jeremy Douglass, Mark C. Marino, Michael Mateas, Casey Reas, Mark Sample, and Noah Vawter, *10 Print CHR$(205.5+RND(1); : GOTO 10* (Cambridge, MA: MIT Press, 2013).

24. Bill Gates reflects on the difficulties that hardware incompatibility posed for software developers during the late 1970s and early 1980s prior to the release of the PC in “A Trend Toward Softness,” *Creative Computing*, November 1984, 121–22.

25. Although there are many books on IBM’s history, few focus specifically on the company after 1980. For the broader corporate histories of IBM that I consulted for background information, see Emerson W. Pugh, *Building IBM: Shaping an Industry and Its Technology* (Cambridge, MA: MIT Press, 1995), and James W. Cortada, *IBM: The Rise and Fall and Reinvention of a Global Icon* (Cambridge, MA: MIT Press, 2019).

26. James Chposky and Ted Leonsis, *Blue Magic: The People, Power and Politics behind the IBM Personal Computer* (New York: Facts on File, 1988), viii.

27. Paul Carroll, *Big Blues: The Unmaking of IBM* (New York: Crown, 1993).

28. Chposky and Leonsis, *Blue Magic*, 8–13.

29. Chposky and Leonsis are light on details here. More information on VisiCalc's origins on the Apple II can be found in Paul Freiberger and Michael Swaine, *Fire in the Valley: The Making of the Personal Computer*, 2nd ed. (New York: McGraw-Hill, 2000), 289–91, and Paul Ceruzzi, *A History of Modern Computing*, 2nd ed. (Cambridge, MA: MIT Press, 2003), 266–68.

30. Carroll, *Big Blues*, 22–23.

31. Carroll, *Big Blues*, 31.

32. David Bunnell, "Boca Diary," *PC Magazine*, April / May 1982, 25.

33. Freiberger and Swaine, *Fire in the Valley*, 346.

34. David J. Bradley, "The Design of the IBM PC," *BYTE*, September 1990, 414.

35. Brian Camenker, "The Making of the IBM PC," *BYTE*, November 1983.

36. Bunnell, "Boca Diary," 26.

37. "Welcome, IBM, to Personal Computing," *BYTE*, December 1975, 90.

38. For a description of the DisplayWriter's design and operation, see Matthew G. Kirschenbaum, *Track Changes: A Literary History of Word Processing* (Cambridge, MA: Harvard University Press, 2016), 59–63.

39. Prices quoted from IBM's online personal computer exhibit, http://www-03.ibm.com/ibm/history/exhibits/pc/pc_1.html, accessed May 27, 2021.

40. Chposky and Leonsis, *Blue Magic*, 87.

41. Apple Computer, Inc. v. Franklin Computer Corp., 714 F.2d 1240 (3d Cir. 1983).

42. Jim Forbes, "Vendors Cook up Clones," *InfoWorld*, April 29, 1985, 69–70.

43. Gregg Williams, "A Closer Look at the IBM Personal Computer," *BYTE*, January 1982, 68.

44. Will Fastie, "The IBM Personal Computer," *Creative Computing*, December 1981, 37.

45. Williams, "A Closer Look at the IBM Personal Computer," 56.

46. Fastie, "The IBM Personal Computer," 38.

47. Jim Edlin, "Confessions of a Convert," *PC Magazine*, February / March 1982, 12.

48. Apple Computer Inc., *Apple II Reference Manual* (Cupertino, CA: Apple Computer Inc., 1978).

49. Jeremy Joan Hewes, "A Glimpse at Two PC Manuals," *PC Magazine*, April / May 1982, 116.

50. Apple Computer Inc., *The Applesoft Tutorial* (Cupertino, CA: Apple Computer Inc., 1979), 14, 17.

51. Apple Computer Inc., *The Applesoft Tutorial*, 17

52. IBM, *Guide to Operations* (Boca Raton, FL: IBM, 1981), iv.

53. Hewes, "A Glimpse at Two PC Manuals," 116.

54. Lawrence J. Curran and Richard S. Shuford, "IBM's Estridge," *BYTE*, November 1983, 88.

55. Curran and Shuford, "IBM's Estridge," 94.

56. "IBM's Don Estridge: Big Blue's Total Strategy," *Personal Computing*, September 1984, 204.

57. "IBM's Don Estridge: Big Blue's Total Strategy," 96–97.

58. Bunnell, "Boca Diary," 24.

59. Curran and Shuford, "IBM's Estridge," 89.

60. Philip D. Estridge, "What Makes A Computer Personal?," *Creative Computing*, November 1984, 94.

61. Chposky and Leonsis, *Blue Magic*, 78–79.

62. IBM, "Right Away, You Can See a Difference," advertisement, *BYTE*, July 1982, 84–85.

63. IBM, "Draw Attention to Yourself," advertisement, *BYTE*, February 1983, 120–21.

64. IBM, "Hello, Information," advertisement, *BYTE*, December 1983, 28–29; IBM, "The New IBM DisplayWrite Series Is Here," advertisement, *BYTE*, September 1984, 6–7.

65. Compatibility was a bit more complicated than simply the ability to use the same parts, but an in-depth examination of those issues is beyond the scope of this project. For more, see James Sumner, "Standards and Compatibility: The Rise of the PC Computing Platform," in *Whose Standards? Standardization, Stability, and Uniformity in the History of Information and Electrical Technologies*, ed. James Sumner and J. N. Gooday (London: Continuum,), 101–27.

66. Charles Rubin and Kevin Strehlo, "Why So Many Computers Look Like the 'IBM Standard,'" *Personal Computing*, March 1984, 55.

67. Dataquest Incorporated, *Consolidated Data Base: U.S. Markets, 1981–1990* (San Jose, CA: Data Quest Incorporated, 1986); Dataquest Incorporated, *Consolidated Data Base: U.S. Markets, 1983–1992* (San Jose, CA: Data Quest Incorporated, 1988).

68. United States Census Bureau, "1980 Overview," https://www.census.gov/history/www/through_the_decades/overview/1980.html, accessed Aug 28, 2018.

69. Peggy Zientara, "When Was The Last Time You Used Your Micro?," *InfoWorld*, November 28, 1983.

70. Sam Edwards, "Why Is Software So Hard to Use?," *BYTE*, December 1983, 128.

71. Edwards, "Why Is Software So Hard to Use?," 128.

72. Charles Rubin, "Some People Should Be Afraid Of Computers," *Personal Computing*, August 1983, 56.

73. Jeffrey Rothfeder, "A New Generation Arrives," *Personal Computing*, February 1983, 48.

74. Paul Heckel, "Designing Software: Make Users' Jargon Work for You," pt. 3, *InfoWorld*, July 26, 1982, 13.

75. Edwards, "Why Is Software So Hard to Use?," 132.

76. Paul Kellam, "Onward to Easier Computing," *Personal Computing*, February 1983, 5.

77. *Pirates of Silicon Valley*, directed by Martyn Burke, Turner Network Television, 1999; *Steve Jobs*, directed by Danny Boyle, Universal Pictures, 2015.

78. Michael Moritz, *Return to the Little Kingdom: Steve Jobs, The Creation of Apple, and How It Changed the World* (New York: Overlook Press, 2009), 247–74.

79. Jef Raskin, "Design Considerations for an Anthrophilic Computer," May 29, 1979, Bitsavers scans of Macintosh Project Selected Papers, http://www.bitsavers.org/pdf/apple/mac/The_Macintosh_Project_Selected_Papers_Feb80.pdf (hereafter cited as Bitsavers MPSP).

80. Raskin, "Design Considerations for an Anthrophilic Computer."
81. David H. Ahl, "The First Decade of Personal Computing," *Creative Computing*, November 1984, 40–44.
82. Raskin, "Design Considerations for an Anthrophilic Computer."
83. Raskin, "Design Considerations for an Anthrophilic Computer."
84. Jef Raskin, "Reply to Jobs, and Personal Motivation," October 2, 1979, Bitsavers MPSP.
85. Raskin, "Reply to Jobs, and Personal Motivation."
86. Jef Raskin, "How Can We Make Computers Truly Personal?," February 10, 1980, Bitsavers MPSP.
87. Raskin, "How Can We Make Computers Truly Personal?."
88. Raskin, "How Can We Make Computers Truly Personal?."
89. Walter Isaacson, *Steve Jobs* (New York: Simon and Schuster, 2011), 93.
90. Gregg Williams, "The Lisa Computer System," *BYTE*, February 1983; Melvin Merkin, "In Love with Lisa," *Creative Computing*, October 1983. The front page of a brochure for the system reads in large lettering "Inventing the Personal Computer. Again." In the pages that follow, the machine is repeatedly referred to as a "personal computer," although at times the label is qualified as one specifically "for the office." See Apple Computer, Inc., "Apple Invents the Personal Computer. Again," advertisement, *Personal Computing*, June 1983, 17–25.
91. Michael Rogers, "The Birth of the LISA," *Personal Computing*, February 1983, 89.
92. Thomas Haigh, "Inventing Information Systems: The Systems Men and the Computer, 1950–1968," *Business History Review* 75, no.1 (2001): 15–61.
93. Wang Laboratories, "PCS-II: Wang's New Computer System," brochure, June 1977, Wang2200.org scans of Wang Laboratories documentation, https://www.wang2200.org/docs/.
94. Wang Laboratories, "Wang's 2200 PCS-III: Your Personal Computer System," brochure, March 1980, https://www.wang2200.org/docs/; Wang Laboratories, "The Wang 220 LVP: The Computer That Means Small Business," brochure, July 1980, https://www.wang2200.org/docs/.
95. Wang Laboratories, "The Perfect Small Business Partner," advertisement, October 1981, https://www.wang2200.org/docs/.
96. Lori Emerson, *Reading and Writing Interfaces: From the Digital to the Bookbound* (Minneapolis: University of Minnesota Press, 2014), chap 2.
97. David Canfield Smith, Charles Irby, and Ralph Kimball, "Designing the Star User Interface," *BYTE*, April 1982, 274.
98. Emerson, *Reading Writing Interfaces*, chap. 2.
99. Apple Computer, Inc., "Apple Invents the Personal Computer. Again," 21.
100. Deborah Wise, "Cautious Response to the Lisa's Market Potential," *InfoWorld*, February 21, 1983, 17.
101. Wise, "Cautious Response to the Lisa's Market Potential," 16–17; John Barry, "Apple Computer Architects Natural Interface for Knowledge Workers," *InfoWorld*, February 28, 1983, 47.
102. Apple Computer, Inc., "Apple Invents the Personal Computer. Again."

103. Apple Computer, Inc., *Lisa Owner's Guide* (Cupertino, CA: Apple Computer, Inc., 1983), B2–B3; Apple Computer, Inc., *Workshop User's Guide for the Lisa* (Cupertino, CA: Apple Computer, Inc., 1983), chap 1.

104. John J. Anderson, "Apple Macintosh: Cutting through the Ballyhoo," *Creative Computing*, July 1984, 22.

105. Gregg Williams, "The Apple Macintosh Computer," *BYTE*, February 1984, 52.

106. Williams, "The Apple Macintosh Computer," 52.

107. Anderson, "Apple Macintosh," 25.

108. Apple Computer, Inc., "Welcome, IBM. Seriously," advertisement, *Wall Street Journal*, August 12, 1981; Owen W. Linzmayer, *Apple Confidential 2.0: The Definitive History of the World's Most Colorful Company* (San Francisco: No Starch Press, 2004), 95–96.

109. "Steve Jobs on the Future of Apple Computer," *Personal Computing*, April 1984, 248.

110. "Steve Jobs on the Future of Apple Computer," 243.

111. Isaacson, *Steve Jobs*, 138.

112. Isaacson, *Steve Jobs*, 143.

113. "Steve Jobs on the Future of Apple Computer," 242.

114. "Steve Jobs on the Future of Apple Computer," 248.

115. All quotations in this paragraph and the next are drawn from Apple Computer, Inc., "Introducing Macintosh," brochure, December 1983, Archive.org, https://archive.org/details/MAC_128K_BROCHURE_1983.

116. Charles Rubin, "Macintosh: Apple's Powerful Computer," *Personal Computing*, February 1984, 56.

117. Rubin, "Macintosh: Apple's Powerful Computer," 58–59.

118. Anderson, "Apple Macintosh: Cutting Through the Ballyhoo," 19.

119. Williams, "The Apple Macintosh Computer," 54.

120. Lev Manovich, *Software Takes Command* (New York: Bloomsbury, 2013), 207–8.

121. Sarah Inman and David Ribes, "'Beautiful Seams': Strategic Revelations and Concealments," in *Proceedings of the 2019 CHI Conference on Human Factors in Computing Systems* (New York: Association for Computing Machinery, 2019).

Chapter 5. The Human Factor

1. Martin Campbell-Kelly, *From Airline Reservations to Sonic the Hedgehog: A History of the Software Industry* (Cambridge, MA: MIT Press, 2003), chap. 1.

2. Elizabeth R. Petrick, "A Historiography of Human-Computer Interaction," *IEEE Annals of the History of Computing* 42, no. 4 (2020): 9.

3. John M. Carroll, "Human Computer Interaction: Psychology as a Science of Design," *Annual Review of Psychology* 48 (1997): 61–83; Johnathan Grudin, "A Moving Target: The Evolution of Human-Computer Interaction," in *The Human-Computer Interaction Handbook: Fundamentals, Evolving Technologies, and Emerging Applications*, ed. Julie A. Jacko (Boca Raton, FL: CRC Press, 2012), xxvii–lxi; Brian Shackel, "Human-Computer Interaction—Whence and Wither?," *Journal of American Society for Information Science* 48, nos. 5–6 (1997): 970–86.

4. This emphasis is visible both in the citations of published research as well in the primary sources included in digital media studies textbooks. See, for example, *The New*

Media Reader, ed. Noah Wardrip-Fruin and Nick Montfort (Cambridge, MA: MIT Press, 2003), *New Media, Old Media: A History and Theory Reader*, ed. Wendy Hui Kyong Chun and Thomas Keenan (New York: Routledge, 2006), Wendy Hui Kyong Chun, *Programmed Visions: Software and Memory* (Cambridge, MA: MIT Press, 2010), Lori Emerson, *Reading Writing Interfaces: From the Digital to the Bookbound* (Minneapolis: University of Minnesota Press, 2014), and Lev Manovich, *Software Takes Command* (New York: Bloomsbury, 2013).

5. Richard Coyne, *Designing Information Technology in the Postmodern Age: From Method to Metaphor* (Cambridge, MA: MIT Press, 1995), 33.

6. Geoff Cooper and John Bowers, "Representing the User: Notes on the Disciplinary Rhetoric of Human-Computer Interaction," *The Social and Interactional Dimensions of Human-Computer Interfaces*, ed. Peter J. Thomas (New York: Cambridge University Press, 1995), 48–66.

7. For a discussion about a similar misalignment between the political motivations within more recent theories of evaluative usability and the observed practice of user-experience testing, see Lillie Ruth Jenkins, "Designing Systems That Make Sense: What Designers Say about Their Communication with Users during the Usability Testing Cycle" (PhD diss., Ohio State University, 2004), 1–16, http://rave.ohiolink.edu/etdc/view?acc_num=osu1086193345.

8. Brian Shackel, "Ergonomics for a Computer," *Design* 120 (1959): 36–39; Brian Shackel, "Ergonomics in the Detail of a Large Digital Computer Console," *Ergonomics* 5, no. 1 (1962): 229–41.

9. Grudin, "A Moving Target," xxviii; Lillian Gilbreth, *The Psychology of Management: The Function of the Mind in Determining Teaching and Installing Methods of Least Waste* (New York: Sturgis and Walton, 1914); Frederick Winslow Taylor, *The Principles of Scientific Management* (New York: Harper, 1911).

10. Henry Dreyfuss, *Designing for People* (New York: Simon and Schuster, 1955).

11. For a history of nondiscretionary users, see Nathan Ensmenger, *The Computer Boys Take Over: Computers, Programmers, and the Politics of Technical Expertise* (Cambridge, MA: MIT Press, 2010).

12. George A. Miller, Eugene Galanter, and Karl H. Pribram, *Plans and the Structure of Behavior* (New York: Holt, Rinehart and Winston, 1960); Ulrich Neisser, *Cognitive Psychology* (Englewood Cliffs, NJ: Prentice-Hall, 1976); Allen Newell and Herbert A. Simon, *Human Problem Solving* (Englewood Cliffs, NJ: Prentice-Hall, 1972).

13. This general description is adapted from Newell and Simon, *Human Problem Solving.*

14. Stuart K. Card, Thomas P. Moran, and Allen Newell, *The Psychology of Human-Computer Interaction* (New York: Lawrence Erlbaum Associates, 1983), 16.

15. Card, Moran, and Newell, *The Psychology of Human-Computer Interaction*, 11–12.

16. Card, Moran, and Newell, *The Psychology of Human-Computer Interaction*, 23–97.

17. Card, Moran, and Newell, *The Psychology of Human-Computer Interaction*, 419.

18. Ben Shneiderman, *Software Psychology: Human Factors in Computer and Information Systems* (Cambridge, MA: Winthrop Publishers, 1980), 271.

19. Shneiderman, *Software Psychology*, 272.

20. Shneiderman, *Software Psychology*, 275–77.

21. Ben Shneiderman, "The Future of Interactive Systems and the Emergence of Direct Manipulation," *Behaviour and Information Technology* 1, no. 3 (1982): 237.

22. Shneiderman, "The Future of Interactive Systems and the Emergence of Direct Manipulation," 251.

23. Shneiderman, "The Future of Interactive Systems and the Emergence of Direct Manipulation," 250.

24. Shneiderman, "The Future of Interactive Systems and the Emergence of Direct Manipulation," 249.

25. Shneiderman, "The Future of Interactive Systems and the Emergence of Direct Manipulation," 251–52.

26. Ben Shneiderman and Richard Mayer, "Syntactic / Semantic Interactions in Programmer Behavior: A Model and Experimental Results," *International Journal of Computer and Information Systems* 8, no. 3 (1979): 219–20.

27. Shneiderman and Richard Mayer, "Syntactic / Semantic Interactions in Programmer Behavior," 224–29.

28. Shneiderman, "The Future of Interactive Systems and the Emergence of Direct Manipulation," 253.

29. Shneiderman and Mayer, "Syntactic / Semantic Interactions in Programmer Behavior," 222–24.

30. Ben Shneiderman, "Direct Manipulation: A Step Beyond Programming Languages," *Computer* 8, no. 16 (1983): 64.

31. Donald A. Norman, *Memory and Attention: An Introduction to Human Information Processing* (New York: Wiley, 1969).

32. Donald A. Norman, "Twelve Issues for Cognitive Science." *Cognitive Science* 4, no. 1 (1980): 1–2.

33. Norman, "Twelve Issues for Cognitive Science," 28–29

34. Donald A. Norman, "Cognitive Engineering," in *User-Centered System Design: New Perspectives in Human-Computer Interaction*, ed. Donald A. Norman and Stephen W. Draper (New York: Lawrence Erlbaum and Associates, 1986), 32.

35. Norman, "Cognitive Engineering," 33–40.

36. Norman, "Cognitive Engineering," 38–39.

37. Norman, "Cognitive Engineering," 47.

38. Norman, "Cognitive Engineering," 43.

39. Norman, "Cognitive Engineering," 54.

40. Norman, "Cognitive Engineering," 45.

41. Norman, "Cognitive Engineering," 60–61.

42. Donald A. Norman, *The Design of Everyday Things* (1988; repr., New York: Basic Books, 2002), 177–78.

43. Norman, *The Design of Everyday Things*, 183.

44. Hubert L. Dreyfus, *Alchemy and Artificial Intelligence* (Santa Monica, CA: RAND Corporation, 1965).

45. Hubert L. Dreyfus, *What Computers Can't Do: A Critique of Artificial Reason* (New York: Harper and Row, 1972).

46. Terry Winograd and Fernando Flores, *Understanding Computers and Cognition: A New Foundation for Design* (1986; repr., Reading: Addison-Wesley, 1987), xiii–xiv.

47. Dreyfus, *What Computers Can't Do*, xxxiv.
48. Dreyfus, *What Computers Can't Do*, xxxv.
49. Dreyfus, *What Computers Can't Do*, xvii.
50. Alan Turing, "Computing Machinery and Intelligence," *Mind* 59, no. 236 (1950): 433–60.
51. Dreyfus, *What Computers Can't Do*, 198–99.
52. Dreyfus, *What Computers Can't Do* 201.
53. Dreyfus, *What Computers Can't Do*, 200.
54. Winograd and Flores, *Understanding Computers and Cognition*, 78.
55. Winograd and Flores, *Understanding Computers and Cognition*, 6.
56. Winograd and Flores, *Understanding Computers and Cognition*, 16.
57. Winograd and Flores, *Understanding Computers and Cognition*, 71–2.
58. Winograd and Flores further note that their interpretation of Heidegger is largely based on Dreyfus's commentary on *Being and Time*, which was not published until 1991. Because my goal here is describe Winograd and Flores's understanding of Heidegger's philosophy, my comments in this remainder of this section refer to how they describe phenomenology and hermeneutics in their book rather than on Dreyfus's commentaries published a decade later or the source works by Heidegger. See Winograd and Flores, *Understanding Computing and Cognition*, 32n, Hubert L. Dreyfus, *Being-in-the-World: A Commentary on Heidegger's "Being and Time," Division I* (Cambridge, MA: MIT Press, 1991).
59. Winograd and Flores, *Understanding Computers and Cognition*, 30–36.
60. Winograd and Flores, *Understanding Computers and Cognition*, 165.
61. Winograd and Flores, *Understanding Computers and Cognition*, 164.
62. Winograd and Flores, *Understanding Computers and Cognition* 164–45.
63. Winograd and Flores, *Understanding Computers and Cognition*, 179.
64. See, for example, Donald A. Norman and Stephen W. Draper, "The Interface Experience," in *User Centered System Design*, ed. Donald A. Norman and Stephen W. Draper (Hillsdale, NJ: Lawrence Erlbaum Associates, 1986), 63–64.
65. My notes in this section refer to the first edition of Laurel's book. In the second edition, published in 2014, she has substantially reorganized the book and has updated her discussion to incorporate more recent research on HCI. Given that my focus is on how usability was understood during the 1980s, I have chosen to limit my examination of her work to the book's first edition.
66. Brenda Laurel, *Computers as Theatre* (Reading, MA: Addison-Wesley, 1991), xix.
67. Laurel, *Computers as Theatre*, xvii.
68. Laurel, *Computers as Theatre*, xix.
69. Laurel, *Computers as Theatre*, 6.
70. Laurel, *Computers as Theatre*, 15.
71. Laurel, *Computers as Theatre*, 17.
72. Laurel, *Computers as Theatre*, 16–17.
73. Laurel, *Computers as Theatre*, 101.
74. Laurel, *Computers as Theatre*, 105.
75. Lucy A. Suchman, *Plans and Situated Actions: The Problem of Human-Machine Communication* (New York: Cambridge University Press, 1987), 57.

76. Charles S. Peirce, "On the Algebra of Logic: A Contribution to the Philosophy of Notation" *American Journal of Mathematics* 7, no. 2 (1885): 180–96; Harold Garfinkel, *Studies in Ethnomethodology* (Englewood Cliffs, NJ: Prentice-Hall, 1967).

77. Suchman, *Plans and Situated Actions*, 63.

78. Suchman discusses the political implications of this boundary in "Working Relations of Technology Production and Use," *Computer Supported Cooperative Work* 2, nos. 1–2 (1994): 21–39.

79. Suchman, *Plans and Situated Actions*, 53.

80. Suchman, *Plans and Situated Actions*, 52.

81. Suchman, *Plans and Situated Actions*, 180.

82. Suchman, *Plans and Situated Actions*, 185.

83. Suchman, *Plans and Situated Actions*, 189.

84. Suchman, "Working Relations of Technology Production and Use," 23–24.

85. Suchman, "Working Relations of Technology Production and Use," 25–27.

86. Lucy Suchman, *Human-Machine Reconfigurations: Plans and Situated Actions*, 2nd edition. (New York: Cambridge University Press, 2007), 182n.

87. *Silicon Valley*, season 1, episode 7, "Proof of Concept," directed by Mike Judge, aired May 18, 2014, HBO.

88. Jeffrey Bardzell and Shaowen Bardzell, *Humanistic HCI* (San Rafael, CA: Morgan and Claypool, 2015), xix.

89. Bardzell and Bardzell, chap 3.

90. Sasha Constanza-Chock, *Design Justice* (Cambridge, MA: MIT Press, 2020); Natasha N. Jones, "Narrative Inquiry in Human-Centered Design: Examining Silence and Voice to Promote Social Justice in Design Scenarios," *Journal of Technical Writing and Communication* 46, no. 4 (2016): 471–92.

91. Toni Robertson and Jesper Simonsen, "Participatory Design: An Introduction," in *Routledge International Handbook of Participatory Design*, ed. Jesper Simonsen and Toni Robertson (New York: Routledge, 2013), 2.

92. Robertson and Simonsen, "Participatory Design," 3–8.

93. Constanza-Chock, *Design Justice*, 110–12.

Coda. Imagining an Unfriendly Future

1. Paul N. Edwards, "Infrastructure and Modernity: Force, Time, and Social Organization in the History of Sociotechnical Systems," in *Modernity and Technology*, ed. Thomas J. Misa, Philip Brey, and Andrew Feenberg (Cambridge: MIT Press, 2003), 185–226.

2. Annette Vee, *Coding Literacy: How Computer Programming Is Changing Writing* (Cambridge, MA: MIT Press, 2017), 27–34.

3. Apple has historically insisted that the consistency necessary to promote "aesthetic integrity" across all software accessed through its platforms is driven by a need to meet "users' expectations." See Apple Computer, Inc., *Macintosh Human Interface Guidelines* (Reading, MA: Addison-Wesley, 1992), 7–11.

4. Lilly Irani, "Hackathons and the Making of Entrepreneurial Citizenship," *Science, Technology, & Human Values* 40, no. 5 (2015): 799–824.

5. Ruha Benjamin, *Race After Technology: Abolitionist Tools for the New Jim Code* (Medford, MA: Polity, 2019), 28–9.

6. M. Remi Yergeau, Elizabeth Brewer, Stephanie L. Kerschbaum, Sushil K. Oswal, Margaret Price, Cynthia L. Selfe, Michael J. Salvo, and Franny Howes, "Modality in Motion: Disability & Kairotic Spaces," *Kairos* 18, no. 1 (2013): http://kairos.technorhetoric.net/18.1/coverweb/yergeau-et-al/pages/access.html.

7. Shoshana Zuboff, *The Age of Surveillance Capitalism: The Fight for a Human Future at the New Frontier of Power* (New York: Public Affairs, 2019), 164–65.

8. "Workplace Analytics: Productivity and Wellbeing," Microsoft Office 365, accessed January 23, 2021, https://insights.office.com/productivityandwellbeing/.

9. Nathan Tenhundfeld, Mustafa Demir, and Ewart J. de Visser, "An Argument for Trust Assessment in Human-Machine Interaction: Overview and Call for Integration," *OSF Preprints*, January 11, 2021, doi:10.31219/osf.io/j47df.

10. Geoffrey A. Fowler, "iPhone App Privacy Labels Are a Great Idea, Except When Apple Lets Them Deceive," Washington Post, January 29, 2021, https://www.washingtonpost.com/technology/2021/01/29/apple-privacy-nutrition-label/.

11. Joseph Cox, "Leaked Location Data Shows Another Muslim Prayer App Tracking Users," *Vice*, January 11, 2021, https://www.vice.com/en/article/xgz4n3/muslim-app-location-data-salaat-first.

12. Zoe Schiffer, "Period Tracking App Settles Charges It Lied to Users about Privacy," *The Verge*, January 13, 2021, https://www.theverge.com/2021/1/13/22229303/flo-period-tracking-app-privacy-health-data-facebook-google.

13. William Gibson, *Neuromancer* (New York: Ace Books, 1984).

14. Steve Krug, *"Don't Make Me Think": A Common Sense Approach to Web Usability*, 2nd ed. (Berkeley, CA: New Riders, 2006).

15. Nicholas Thompson, "Our Minds Have Been Hijacked by Our Phones: Tristan Harris Wants to Rescue Them," *Wired*, July 26, 2017, https://www.wired.com/story/our-minds-have-been-hijacked-by-our-phones-tristan-harris-wants-to-rescue-them/.

16. Chris Gillard and David Golumbia, "There Are No Guardrails on Our Privacy Dystopia," *Vice*, March 9, 2018, https://www.vice.com/en/article/zmwaee/there-are-no-guardrails-on-our-privacy-dystopia.

17. Lucy Suchman, "Working Relations of Technology Production and Use," *Computer Supported Cooperative Work* 2, nos. 1–2 (1993): 23–24.

18. Joel Franey, "Why Does Skyrim Feel So Strangely Old These Days?," *USGamer*, August 11, 2020, https://www.usgamer.net/articles/skyrim-nostalgia-retrospective.

19. Jenny Chan, Mark Selden, and Pun Ngai, *Dying for an iPhone: Apple, Foxconn, and the Lives of China's Workers* (Chicago: Haymarket Books, 2020).

20. Caroline O'Donovan, "Google Employees Are Organizing to Protest the Company's Secret, Censored Search Engine for China," *Buzzfeed*, August 17, 2018, https://www.buzzfeednews.com/article/carolineodonovan/google-dragonfly-maven-employee-protest-demands.

21. Nathan L. Ensmenger, *The Computer Boys Take Over: Computers, Programmers, and the Politics of Technical Expertise* (Cambridge, MA: MIT Press, 2012), chap 8.

Index